Annals of Mathematics Studies

Number 117

RADICALLY ELEMENTARY PROBABILITY THEORY

BY

EDWARD NELSON

PRINCETON UNIVERSITY PRESS

———

PRINCETON, NEW JERSEY

1987

Clothbound editions of Princeton University Press
books are printed on acid-free paper, and binding
materials are chosen for strength and durability. Pa-
perbacks, while satisfactory for personal collections,
are not usually suitable for library rebinding

ISBN 0-691-08473-4 (cloth)
ISBN 0-691-08474-2 (paper)

Printed in the United States of America
by Princeton University Press, 41 William Street
Princeton, New Jersey

☆

Library of Congress Cataloging in Publication data will
be found on the last printed page of this book

Table of contents

Preface   vii
Acknowledgments   ix
 1. Random variables   3
 2. Algebras of random variables   6
 3. Stochastic processes   10
 4. External concepts   12
 5. Infinitesimals   16
 6. External analogues of internal notions   20
 7. Properties that hold almost everywhere   25
 8. L^1 random variables   30
 9. The decomposition of a stochastic process   33
10. The total variation of a process   37
11. Convergence of martingales   41
12. Fluctuations of martingales   48
13. Discontinuities of martingales   53
14. The Lindeberg condition   57
15. The maximum of a martingale   61
16. The law of large numbers   63
17. Nearly equivalent stochastic processes   72
18. The de Moivre-Laplace-Lindeberg-Feller-Wiener-
    Lévy-Doob-Erdös-Kac-Donsker-Prokhorov theorem   75
Appendix   80
Index   95

Preface

More than any other branch of mathematics, probability theory has developed in conjunction with its applications. This was true in the beginning, when Pascal and Fermat tackled the problem of finding a fair way to divide the stakes in a game of chance, and it continues to be true today, when the most exciting work in probability theory is being done by physicists working on statistical mechanics.

The foundations of probability theory were laid just over fifty years ago, by Kolmogorov. I am sure that many other probabilists teaching a beginning graduate course have also had the feeling that these measure-theoretic foundations serve more to salve our mathematical consciences than to provide an incisive tool for the scientist who wishes to apply probability theory.

This work is an attempt to lay new foundations for probability theory, using a tiny bit of nonstandard analysis. The mathematical background required is little more than that which is taught in high school, and it is my hope that it will make deep results from the modern theory of stochastic processes readily available to anyone who can add, multiply, and reason.

What makes this possible is the decision to leave the results in nonstandard form. Nonstandard analysts have a new way of thinking about mathematics, and if it is not translated back into conventional terms then it is seen to be remarkably elementary.

Mathematicians are quite rightly conservative and suspicious of new ideas. They will ask whether the results developed here are as powerful as the conventional results, and whether it is worth their while to learn nonstandard methods. These questions are addressed in an appendix, which assumes a much greater level of mathematical knowledge than does the main text. But I want to emphasize that the main text stands on its own.

Acknowledgments

I am grateful to Eric Carlen, Mai Gehrke, Klaus Kaiser, and Brian White for helpful comments, and to Pierre Cartier for a critical reading of the manuscript and the correction of many errors, together with helpful suggestions. This work was partially supported by the National Science Foundation. I thank Morgan Phillips for the six illustrations that LaTeX did not draw.

Radically Elementary Probability Theory

Chapter 1

Random variables

Here are some of the basic definitions and inequalities of probability theory, in the context of a finite probability space.

A *finite probability space* is a finite set Ω and a strictly positive function pr on Ω such that $\sum \text{pr}(\omega) = 1$. Then a *random variable* on Ω is a function $x: \Omega \to \mathbf{R}$, where $\mathbf{R}$ is the real numbers. The *expectation* or *mean* of a random variable x is

$$\mathbf{E}x = \sum x(\omega)\text{pr}(\omega).$$

An *event* is a subset A of Ω, and the *probability* of an event A is

$$\Pr A = \sum_{\omega \in A} \text{pr}(\omega).$$

If A is an event, we define a random variable χ_A, called its *indicator function*, by $\chi_A(\omega) = 1$ if $\omega \in A$ and $\chi_A(\omega) = 0$ if $\omega \notin A$. Then $\Pr A = \mathbf{E}\chi_A$. Also, we define A^c to be the complementary event $A^c = \Omega \setminus A$ of all ω in Ω that are not in A.

The set $\mathbf{R}^\Omega$ of all random variables on Ω is an n-dimensional vector space, where n is the number of points in Ω. Consider the expression $\mathbf{E}xy$, where x and y are any two random variables. Then $\mathbf{E}xy = \mathbf{E}yx$, $\mathbf{E}xy$ is linear in x and y, and $\mathbf{E}xx > 0$ unless $x = 0$. Thus $\mathbf{E}xy$ has all of the properties of the inner product on n-dimensional Euclidean space. The Euclidean norm $\sqrt{\mathbf{E}x^2}$ of the random variable x is denoted by $\|x\|_2$.

The expectation is a linear function on $\mathbf{R}^\Omega$, so the random variables of mean 0 form a hyperplane. The orthogonal complement of this hyperplane is the one-dimensional subspace of constant random variables. We identify the constant random variable whose value is λ with the number λ. With this identification, $x \mapsto \mathbf{E}x$ is the orthogonal projection onto the constant random variables and $x \mapsto x - \mathbf{E}x$ is the orthogonal projection onto the

3

random variables of mean 0. We call $\mathrm{Var}\, x = \mathbf{E}(x - \mathbf{E}x)^2$ the *variance* of x, $\sqrt{\mathrm{Var}\, x}$ the *standard deviation* of x, $\mathbf{E}(x - \mathbf{E}x)(y - \mathbf{E}y)$ the *covariance* of x and y, and

$$\frac{\mathbf{E}(x - \mathbf{E}x)(y - \mathbf{E}y)}{\sqrt{\mathrm{Var}\, x}\,\sqrt{\mathrm{Var}\, y}}$$

the *correlation coefficient* of x and y. Thus if x and y have mean 0, the variance of x is the square $\|x\|_2^2$ of its Euclidean norm, the standard deviation of x is its Euclidean norm $\|x\|_2$, the covariance of x and y is their inner product, and the correlation coefficient of x and y is the cosine of the angle between them.

Other norms on random variables are frequently useful. For $1 \le p < \infty$ let $\|x\|_p = (\mathbf{E}|x|^p)^{1/p}$, and let $\|x\|_\infty = \max |x(\omega)|$. Clearly $\|x\|_p \le \|x\|_\infty$, and if ω_0 is a point at which $|x|$ attains its maximum, then

$$\|x\|_p \ge (|x(\omega_0)|^p \mathrm{pr}(\omega_0))^{1/p} \to \|x\|_\infty$$

as $p \to \infty$, so that $\|x\|_p \to \|x\|_\infty$ as $p \to \infty$.

For $1 < p < \infty$, let p', called the *conjugate exponent* to p, be defined by $p' = p/(p - 1)$, so that

$$\frac{1}{p} + \frac{1}{p'} = 1.$$

Hölder's inequality asserts that

$$|\mathbf{E}xy| \le \|x\|_p \|y\|_{p'}. \tag{1.1}$$

If x or y is 0, this is trivially true. Otherwise, we can assume that we have $\|x\|_p = \|y\|_{p'} = 1$ after replacing x by $x/\|x\|_p$ and y by $y/\|y\|_{p'}$. Then $\mathbf{E}|x|^p = \mathbf{E}|y|^{p'} = 1$ and we want to show that $|\mathbf{E}xy| \le 1$. Since $|\mathbf{E}xy| \le \mathbf{E}|xy|$, this will follow if we can show that $|xy|$ is less than a convex combination of $|x|^p$ and $|y|^{p'}$. Taking the obvious convex combination, we need only show that

$$|xy| \le \frac{1}{p}|x|^p + \frac{1}{p'}|y|^{p'}.$$

To see this, take logarithms. By the concavity of the logarithm function, the logarithm of the right hand side is greater than

$$\frac{1}{p} \log |x|^p + \frac{1}{p'} \log |y|^{p'} = \log |xy|,$$

which concludes the proof of Hölder's inequality.

Keeping the normalizations $\|x\|_p = \|y\|_{p'} = 1$, we see that we have strict inequality in (1.1) unless $|\mathbf{E}xy| = \mathbf{E}|xy|$—that is, unless x and y have the same sign—and unless $|x|^p = |y|^{p'}$; but if $x = \operatorname{sgn} y\, |y|^{p'/p}$, then $\mathbf{E}xy = 1$. Consequently, for all random variables x we have

$$\|x\|_p = \max_{\|y\|_{p'}=1} |\mathbf{E}xy|. \tag{1.2}$$

An immediate consequence of (1.2) is *Minkowski's inequality*

$$\|x + z\|_p \le \|x\|_p + \|z\|_p. \tag{1.3}$$

If we let $1' = \infty$ and $\infty' = 1$, then (1.1), (1.2), and (1.3) hold for all p with $1 \le p \le \infty$.

Let f be a convex function. By definition this means that

$$f\!\left(\sum x(\omega)\mathrm{pr}(\omega)\right) \le \sum f(x(\omega))\mathrm{pr}(\omega);$$

that is,

$$f(\mathbf{E}x) \le \mathbf{E}f(x), \tag{1.4}$$

which is *Jensen's inequality*. If we apply this to the convex function $f(x) = |x|^p$, where $1 \le p < \infty$, we obtain $|\mathbf{E}x|^p \le \mathbf{E}|x|^p$. Applied to $|x|$ this gives $\|x\|_1 \le \|x\|_p$, and applied to $|x|^r$, where $1 \le r < \infty$, this gives $\|x\|_r \le \|x\|_{rp}$, so that $\|x\|_p$ is an increasing function of p for $1 \le p \le \infty$.

Let f be a positive function. Then, for $\lambda > 0$,

$$\mathbf{E}f(x) = \sum f(x(\omega))\mathrm{pr}(\omega) \ge \sum_{\omega \in \{f(x) \ge \lambda\}} f(x(\omega))\mathrm{pr}(\omega) \ge \lambda \Pr\{f(x) \ge \lambda\},$$

so that

$$\Pr\{f(x) \ge \lambda\} \le \frac{\mathbf{E}f(x)}{\lambda}. \tag{1.5}$$

(Here $\{f(x) \ge \lambda\}$ is an abbreviation for $\{\omega \in \Omega : f(x(\omega)) \ge \lambda\}$; such abbreviations are customary in probability theory.) In particular, for $\lambda > 0$ and $p > 0$ we have $\{|x| \ge \lambda\} = \{|x|^p \ge \lambda^p\}$, and so by (1.5) we have

$$\Pr\{|x| \ge \lambda\} \le \frac{\mathbf{E}|x|^p}{\lambda^p}. \tag{1.6}$$

This is *Chebyshev's inequality*.

Chapter 2

Algebras of random variables

The set $\mathbf{R}^\Omega$ of all random variables on Ω is not only a vector space, it is an algebra. By an *algebra* $\mathcal{A}$ of random variables we will always mean a subalgebra of $\mathbf{R}^\Omega$ containing the constants; that is, $\mathcal{A}$ is a set of random variables containing the constants and such that whenever x and y are in $\mathcal{A}$, then $x + y$ and xy are in $\mathcal{A}$.

The structure of an algebra $\mathcal{A}$ is very simple. By an *atom* of $\mathcal{A}$ we mean a maximal event A such that each random variable in $\mathcal{A}$ is constant on A. Thus Ω is partitioned into atoms—that is, Ω is the union of the atoms and different atoms are disjoint. If A is an atom and $\omega \notin A$, then by definition there is an x in $\mathcal{A}$ such that $x(A) \neq x(\omega)$. Let $x_\omega = (x - x(\omega))/(x(A) - x(\omega))$. Then $x_\omega \in \mathcal{A}$ and $x_\omega(A) = 1$, $x_\omega(\omega) = 0$. Consequently, the indicator function χ_A is in $\mathcal{A}$, since

$$\chi_A = \prod_{\omega \notin A} x_\omega.$$

Thus $\mathcal{A}$ consists of all random variables that are constant on the atoms of $\mathcal{A}$. Conversely, given an arbitrary partition of Ω, the set of all random variables that are constant on each event in the partition is an algebra of random variables.

Notice that an algebra $\mathcal{A}$ of random variables contains arbitrary functions of its elements: if $f : \mathbf{R}^n \to \mathbf{R}$ and $x_1, \ldots, x_n$ are in $\mathcal{A}$, then $f(x_1, \ldots, x_n)$ is in $\mathcal{A}$.

As an example, let Ω be the set of all pairs $\langle i, j \rangle$ with $1 \leq i, j \leq 6$ and $\mathrm{pr}(\langle i, j \rangle) = 1/36$ for all $\langle i, j \rangle$ in Ω. This is a model for throwing a pair of dice. Let $x(\langle i, j \rangle) = i$, $y(\langle i, j \rangle) = j$, and $z(\langle i, j \rangle) = i + j$. Let $\mathcal{A}$ be the smallest algebra containing z. The atoms of $\mathcal{A}$ are indicated in Fig. 2.1. There are 11 atoms which we denote by $A_2, \ldots, A_{12}$, with the subscript denoting the value of z on the atom.

Let A be an algebra of random variables. If A is an atom of A, then A is itself a finite probability space with respect to pr_A defined for all ω in A by

$$\text{pr}_A(\omega) = \frac{1}{\Pr A}\text{pr}(\omega).$$

This means that *every construct or theorem of probability theory can be relativized to any algebra A of random variables.* In this relativization, elements of A play the role of constants.

For example, corresponding to the expectation $\mathbf{E}x$ of a random variable x we define the *conditional expectation* or *conditional mean* $\mathbf{E}_A x$ to be the element of A that on each atom A is the expectation of x with respect to pr_A. Thus if A_ω is the atom containing ω, then

$$\mathbf{E}_A x(\omega) = \frac{1}{\Pr A_\omega} \sum_{\eta \in A_\omega} x(\eta)\text{pr}(\eta).$$

The expectation is linear; that is,

$$\mathbf{E}(\lambda_1 x_1 + \lambda_2 x_2) = \lambda_1 \mathbf{E}x_1 + \lambda_2 \mathbf{E}x_2; \quad \lambda_1, \lambda_2 \in \mathbf{R}.$$

The conditional expectation is A-linear; that is,

$$\mathbf{E}_A(y_1 x_1 + y_2 x_2) = y_1 \mathbf{E}_A x_1 + y_2 \mathbf{E}_A x_2; \quad y_1, y_2 \in A.$$

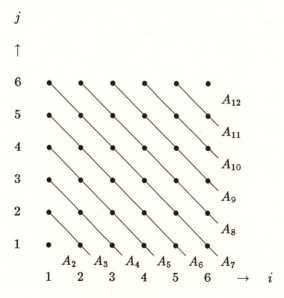

Figure 2.1: Dice

The expectation preserves constants; that is, $\mathbf{E}\lambda = \lambda$ if λ is a constant. The conditional expectation preserves elements of $\mathcal{A}$; that is, $\mathbf{E}_{\mathcal{A}}y = y$ if y is in $\mathcal{A}$. The expectation is the orthogonal projection onto the constants; that is, $\mathbf{E}(x - \mathbf{E}x)^2 \leq \mathbf{E}(x - \lambda)^2$ for all constants λ. By relativization, $\mathbf{E}_{\mathcal{A}}(x - \mathbf{E}_{\mathcal{A}}x)^2 \leq \mathbf{E}_{\mathcal{A}}(x - y)^2$ for all y in $\mathcal{A}$. Notice that $\mathbf{E}_{\mathcal{B}}\mathbf{E}_{\mathcal{A}} = \mathbf{E}_{\mathcal{B}}$ if $\mathcal{B} \subseteq \mathcal{A}$. In particular, $\mathbf{E}\mathbf{E}_{\mathcal{A}} = \mathbf{E}$ since $\mathbf{E}$ is the conditional expectation with respect to the trivial algebra consisting of the constants. Therefore $\mathbf{E}(x - \mathbf{E}_{\mathcal{A}}x)^2 \leq \mathbf{E}(x - y)^2$ for all y in $\mathcal{A}$, so that $\mathbf{E}_{\mathcal{A}}$ is the orthogonal projection onto $\mathcal{A}$.

Another notation for $\mathbf{E}_{\mathcal{A}}x$ is $\mathbf{E}\{x|\mathcal{A}\}$, and we use $\mathbf{E}\{x|x_1,\ldots,x_n\}$ for the conditional expectation of x with respect to the algebra generated by $x_1,\ldots,x_n$. In the example of Fig. 2.1, $\mathbf{E}\{x|z\} = \mathbf{E}\{y|z\}$ by symmetry, so that $\mathbf{E}\{x|z\} = \frac{1}{2}\mathbf{E}\{z|z\} = \frac{1}{2}z$.

We define the *conditional probability* $\mathrm{Pr}_{\mathcal{A}} B$ of an event B with respect to the algebra $\mathcal{A}$, by relativization, as $\mathrm{Pr}_{\mathcal{A}} B = \mathbf{E}_{\mathcal{A}}\chi_B$. Thus

$$\mathrm{Pr}_{\mathcal{A}}B(\omega) = \frac{\mathrm{Pr}(B \cap A_\omega)}{\mathrm{Pr}\, A_\omega},$$

where A_ω is the atom of $\mathcal{A}$ containing ω. Note that this is a random variable, not (in general) a constant.

The relativization of Hölder's inequality is

$$|\mathbf{E}_{\mathcal{A}}xy| \leq (\mathbf{E}_{\mathcal{A}}|x|^p)^{1/p}(\mathbf{E}_{\mathcal{A}}|y|^{p'})^{1/p'},$$

the relativization of Jensen's inequality is

$$f(\mathbf{E}_{\mathcal{A}}x) \leq \mathbf{E}_{\mathcal{A}}f(x)$$

for f convex, and the relativization of Chebyshev's inequality is

$$\mathrm{Pr}_{\mathcal{A}}\{f(x) \geq y\} \leq \frac{\mathbf{E}_{\mathcal{A}}f(x)}{y}$$

for f positive and $y > 0$ in $\mathcal{A}$. From Jensen's inequality we have

$$|\mathbf{E}_{\mathcal{A}}x|^p \leq \mathbf{E}_{\mathcal{A}}|x|^p,$$

and since $\mathbf{E}\mathbf{E}_{\mathcal{A}} = \mathbf{E}$, this gives $\|\mathbf{E}_{\mathcal{A}}x\|_p \leq \|x\|_p$, valid for $1 \leq p \leq \infty$.

Let $\mathcal{A}$ be an algebra of random variables. We denote the set of all atoms of $\mathcal{A}$ by $\mathrm{at}(\mathcal{A})$. Not only is each element A of $\mathrm{at}(\mathcal{A})$ a finite probability space with respect to pr_A, but $\mathrm{at}(\mathcal{A})$ is itself a finite probability space with respect to

$$\mathrm{pr}'_{\mathcal{A}}(A) = \mathrm{Pr}(A)$$

for A in $\mathrm{at}(\mathcal{A})$. We say that the original probability space $\langle \Omega, \mathrm{pr} \rangle$ is *fibered* over $\langle \mathrm{at}(\mathcal{A}), \mathrm{pr}'_A \rangle$, with *fibers* $\langle A, \mathrm{pr}_A \rangle$. In the example of Fig. 2.1, this can be visualized by rotating the figure 45° clockwise. Expectations with respect to pr'_A are denoted by $\mathbf{E}'_A$, and the probability of a set of atoms is denoted by Pr'_A. Notice that $\mathbf{E}'_A \mathbf{E}_A x = \mathbf{E} x$.

A special case of a fibering is a product. Suppose that $\langle \Omega_1, \mathrm{pr}_1 \rangle$ and $\langle \Omega_2, \mathrm{pr}_2 \rangle$ are finite probability spaces. Then $\Omega_1 \times \Omega_2$ is a finite probability space with respect to $\mathrm{pr}_1 \times \mathrm{pr}_2$, where

$$\mathrm{pr}_1 \times \mathrm{pr}_2(\langle \omega_1, \omega_2 \rangle) = \mathrm{pr}_1(\omega_1) \mathrm{pr}_2(\omega_2).$$

Let $\mathcal{A}_1$ be the algebra of all random variables that are functions of ω_1 alone. Then $\mathrm{at}(\mathcal{A}_1)$ consists of all sets of the form $\{\langle \omega_1, \omega_2 \rangle : \omega_1 = \eta_1\}$, where η_1 is any element of Ω_1.

Chapter 3

Stochastic processes

The word "stochastic" means random, and "process" in this context means function, so that a stochastic process is a function whose values are random variables. Let T be a finite set and let $\langle \Omega, \mathrm{pr} \rangle$ be a finite probability space. A *stochastic process indexed by T and defined over $\langle \Omega, \mathrm{pr} \rangle$* is a function $\xi: T \to \mathbf{R}^\Omega$. By a "stochastic process" we will always mean one that is indexed by a finite set and defined over a finite probability space. We write $\xi(t, \omega)$ for the value of $\xi(t)$ at ω, and we write $\xi(\cdot, \omega)$ for the function $t \mapsto \xi(t, \omega)$. Thus each $\xi(t)$ is a random variable, each $\xi(t, \omega)$ is a real number, and each $\xi(\cdot, \omega)$ is a real-valued function on T, called a *trajectory* or *sample path* of the process.

Let Λ_ξ be the set of all trajectories of the stochastic process ξ. Then Λ_ξ is a finite subset of the finite dimensional vector space $\mathbf{R}^T$ of all functions from T to $\mathbf{R}$. We define the *probability distribution* pr_ξ by

$$\mathrm{pr}_\xi(\lambda) = \Pr\{\xi(t) = \lambda(t) \text{ for all } t \text{ in } T\},$$

for all λ in Λ_ξ. Then $\langle \Lambda_\xi, \mathrm{pr}_\xi \rangle$ is a finite probability space. Two stochastic processes ξ and ξ', indexed by the same finite set T but defined over possibly different finite probability spaces, are called *equivalent* in case $\Lambda_\xi = \Lambda_{\xi'}$ and $\mathrm{pr}_\xi = \mathrm{pr}_{\xi'}$; that is, in case they have the same trajectories with the same probabilities. *Probability theory is concerned only with those properties of a stochastic process that are shared by all equivalent stochastic processes.* Notice that the function sending t into the evaluation map $\lambda \mapsto \lambda(t)$ is a stochastic process defined over $\langle \Lambda_\xi, \mathrm{pr}_\xi \rangle$ that is equivalent to ξ. Thus in studying a stochastic process there is no loss of generality in assuming that it is defined over the space of its trajectories.

If T consists of a single element, then we can identify a stochastic process indexed by T with the corresponding random variable. That is, a random variable is a simple special case of a stochastic process. If x is a random variable, then $\mathrm{pr}_x(\lambda) = \Pr\{x = \lambda\}$, where λ is in

$$\Lambda_x = \{\lambda \in \mathbf{R} : \Pr\{x = \lambda\} \neq 0\},$$

and it is easy to see that

$$\mathbf{E}x = \sum \lambda \, \mathrm{pr}_x(\lambda). \tag{3.1}$$

The right hand side of (3.1) is the expectation of the identity function λ on $\langle \Lambda_x, \mathrm{pr}_x \rangle$. (If (3.1) were not true, then, by the dictum of the preceding paragraph, expectations would be of no concern to probability theory.)

The random variables of a stochastic process ξ are called *independent* in case

$$\mathrm{pr}_\xi(\lambda) = \prod_{t \in T} \mathrm{pr}_{\xi(t)}(\lambda(t))$$

for all λ in Λ_ξ. Suppose for example that $T = \{1, \ldots, \nu\}$ and that x_0 is a random variable on $\langle \Omega, \mathrm{pr} \rangle$. Then $\langle \Omega^\nu, \mathrm{pr}^\nu \rangle$, where

$$\mathrm{pr}^\nu(\omega_1, \ldots, \omega_\nu) = \mathrm{pr}(\omega_1) \cdots \mathrm{pr}(\omega_\nu),$$

is a finite probability space, and the random variables x_n defined by

$$x_n(\omega_1, \ldots, \omega_\nu) = x_0(\omega_n), \quad 1 \leq n \leq \nu,$$

are independent. This stochastic process describes ν repeated independent observations of the given random variable x_0.

If $A_1, \ldots, A_\nu$ are events, they are called *independent* in case their indicator functions are independent. In the dice-throwing example (Fig. 2.1), let A be the event that i is odd, let B be the event that j is odd, and let C be the event that $i + j$ is odd. Then A, B are independent; A, C are independent; B, C are independent; but A, B, C are not independent. The event A tells us nothing about C, the event B tells us nothing about C, but A and B together tell us everything about C. This is the principle on which a good detective story is based.

Chapter 4

External concepts

Let x_0 be a random variable of mean 0 and variance 1. Consider ν independent observations $x_1, \ldots, x_\nu$ of x_0.

> If ν is a large number, then almost surely for all large $n \le \nu$ the average $(x_1 + \cdots + x_n)/n$ is nearly equal to 0.

This is an intuitive statement of the strong law of large numbers. It is not precise because we have not explained what is meant by "large", "almost surely", and "nearly equal".

Here is a sketch of how the strong law of large numbers is formulated in conventional mathematics. One replaces the finite sequence $x_1, \ldots, x_\nu$ by an actually infinite sequence $x_1, x_2, \ldots$ To do this, one must construct the infinite Cartesian product of the initial probability space Ω with itself. Even if Ω is a finite probability space, the infinite Cartesian product will contain an uncountable infinity of points. Each individual point has probability 0. Only certain (measurable) sets are events, and the probability of an event is no longer the sum of the probabilities of the points in it. Only certain (measurable) functions are random variables. The expectation becomes an integral, and only certain (integrable) random variables have expectations. Then the strong law of large numbers becomes the statement that except for an event of probability 0, for all $\varepsilon > 0$ there is an m such that for all $n \ge m$ we have $|(x_1 + \cdots + x_n)/n| \le \varepsilon$.

The approach that we take is different, and has the virtue of remaining within the elementary framework of finite probability spaces. We retain the finite sequence $x_1, \ldots, x_\nu$ but we let ν be nonstandard. By an infinitesimal we mean a real number whose absolute value is smaller than the reciprocal of some nonstandard natural number, and two real numbers are said to be nearly equal in case their difference is infinitesimal. A property holds almost surely in case for all non-infinitesimal positive ε there is an event N with $\Pr N \leq \varepsilon$ such that the property holds on N^c. Then, with "large" as a synonym for nonstandard, the statement of the strong law of large numbers is identical with the intuitive statement given above.

The conventional approach involves an idealization, because one cannot actually complete an infinite number of observations. The second approach also involves an idealization, because one cannot actually complete a nonstandard number of observations. In fact, it is in the nature of mathematics to deal with idealizations. The choice of a formalism must be based on esthetic considerations, such as directness of expression, simplicity, and power. Actually, different formalisms in no way exclude each other, and it can be illuminating to look at familiar material from a fresh point of view.

Let us examine how the notion of a nonstandard number arises. Let $\mathbf{N}$ be the set of all natural numbers $0, 1, 2, \ldots$ The basic property of $\mathbf{N}$ is the *induction theorem*, which asserts that if S is a subset of $\mathbf{N}$ containing 0, and such that whenever n is in S then $n + 1$ is in S, then $S = \mathbf{N}$. Now if $\mathrm{A}(n)$ is any formula of conventional mathematics, such as "n is prime and $n + 2$ is prime" or "$n \geq m$", then we can form the subset $S = \{n \in \mathbf{N} : \mathrm{A}(n)\}$ of all natural numbers n for which $\mathrm{A}(n)$ holds. However, the formula must be a formula of the agreed-upon language for mathematics. Sets are not objects in the real world; they are formal mathematical objects and only exist when the formal rules of mathematics say they exist. For example, it does not make sense to consider $S = \{n \in \mathbf{N} : \mathrm{A}(n)\}$ if $\mathrm{A}(n)$ is "n is not in my opinion enormously large".

From the work of Gödel in the early thirties it emerged that the basic intuitive systems of mathematics, such as $\mathbf{N}$, cannot be completely characterized by any axiom scheme. To explain what this means, let us adjoin to the language of conventional mathematics a new undefined predicate "standard". Then "n is standard" has no meaning within conventional mathematics. We call a formula *internal* in case it does not involve "standard"—that is, in case it is a formula of conventional mathematics—and otherwise we call it *external*. Thus the simplest example of an external formula is "n is standard". Another example of an external formula is "x is infinitesimal", since by definition this means: there exists a nonstandard natural number ν such that $|x| \leq 1/\nu$. *Only internal formulas may be used to form subsets.* (For example, it makes no sense to speak of "the set of all standard natural numbers" or "the set of all infinitesimal real numbers".) We call an abuse of this rule *illegal set formation*.

We make the following assumptions:

1. 0 *is standard,*

2. *for all n in* $\mathbf{N}$, *if n is standard then* $n + 1$ *is standard.*

Then it is impossible to prove that every n in $\mathbf{N}$ is standard. (This does not contradict the induction theorem—it merely shows that it is impossible to prove that there is a subset S of $\mathbf{N}$ such that a natural number n is in S if and only if n is standard.) That is, it is consistent to assume also:

3. *there exists a nonstandard n in* $\mathbf{N}$.

We also assume:

4. *if* $A(0)$ *and if for all standard n whenever* $A(n)$ *then* $A(n + 1)$, *then for all standard n we have* $A(n)$.

In (4), $A(n)$ is any formula, internal or external. This assumption is called *external induction.* It is a complement to ordinary induction, which as we have seen may fail for external formulas. (Of course, ordinary induction continues to hold for ordinary—i.e., internal—formulas. Nothing in conventional mathematics is changed; we are merely constructing a richer language to discuss the same mathematical objects as before.)

Using external induction we can easily prove that *every nonstandard natural number is greater than every standard natural number* (let ν be a nonstandard natural number and in (4) let $A(n)$ be "$\nu \geq n$"), that *the sum of two standard natural numbers is standard* (let m be a standard natural number and let $A(n)$ be "$n + m$ is standard"), and that *the product of two standard natural numbers is standard* (let m be a standard natural number, let $A(n)$ be "nm is standard", and use the fact just proved about sums).

Another assumption that we shall occasionally use is called the *sequence principle.* Let $A(n, x)$ be a formula, internal or external. If for all standard n there is an x such that $A(n, x)$, then, of course, there is an x_0 such that $A(0, x_0)$, an x_1 such that $A(1, x_1)$, an x_2 such that $A(2, x_2)$, and so forth. We assume:

*5. *if for all standard n there is an x such that* $A(n, x)$, *then there is a sequence* $n \mapsto x_n$ *such that for all standard n we have* $A(n, x_n)$.

For an example of the use of the sequence principle, see the proof of Theorem 6.1. Results that use the sequence principle will be starred.

Notice that by (2) there is no smallest nonstandard natural number. We can picture the natural numbers as lying on a tape (Fig. 4.1). The standard natural numbers behave just like the full system $\mathbf{N}$, so far as internal properties are concerned. But $\mathbf{N}$ consists of the standard and the nonstandard natural numbers as well. Notice that we did not start with the left portion of the tape and invent a right portion to be added on.

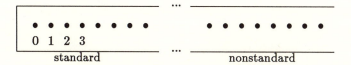

Figure 4.1: The natural numbers

Rather we started with the whole tape and then adjoined a predicate to our language that allows us to distinguish the two portions of the tape. The use of this new predicate "standard" is similar to color on a TV set: the picture is the same, but we see distinctions that we could not make before.

For a long time the incompleteness of axiomatic systems was regarded by mathematicians as unfortunate. It was the genius of Abraham Robinson, in the early sixties, to turn it to good use and show that thanks to it a vast simplification of mathematical reasoning can be achieved.

Chapter 5

Infinitesimals

Now we introduce some useful external notions for the field $\mathbf{R}$ of real numbers.

A real number x is called *infinitesimal* in case $|x| \leq 1/\nu$ for some nonstandard natural number ν. Since a nonstandard ν is bigger than every standard n, it follows that if x is infinitesimal then $|x| \leq 1/n$ for every standard natural number n. I claim that the converse is also true. If $|x| \leq 1/n$ for all n in $\mathbf{N}$, then $x = 0$, and so is infinitesimal; otherwise, let μ be the least natural number such that $|x| > 1/\mu$. Then μ is nonstandard, so $\nu = \mu - 1$ is nonstandard. But $|x| \leq 1/\nu$, so x is infinitesimal.

A real number x is called *limited* in case $|x| \leq n$ for some standard n in $\mathbf{N}$; otherwise x is called *unlimited*. The words "finite" and "infinite" are sometimes used as synonyms for "limited" and "unlimited", respectively, but since they already have internal meanings, their use can lead to confusion, as in "this integral is finite".

If x and y are real numbers we say that:

$x \simeq y$ in case $x - y$ is infinitesimal,

$x \lesssim y$ in case $x \leq y + \alpha$ for some infinitesimal α,

$x \gtrsim y$ in case $y \lesssim x$,

$x \ll y$ in case $x < y$ and $x \not\simeq y$,

$x \gg y$ in case $y \ll x$.

We may read $x \simeq y$ as x is *infinitely close* (or *nearly equal*) to y, $x \lesssim y$ as x is *weakly less than* y, $x \gtrsim y$ as x is *weakly greater than* y, $x \ll y$ as x is *strongly less than* y, $x \gg y$ as x is *strongly greater than* y, and $x \gg 0$ as x is *strongly positive*.

The extended real line $\overline{\mathbf{R}}$ consists of $\mathbf{R}$ together with two other points, $-\infty$ and ∞. We write $-\infty < x < \infty$ for all x in $\mathbf{R}$. We define $x \simeq \infty$ to mean that x is positive and unlimited, $x \ll \infty$ (or $\infty \gg x$) to mean that

$x \not\simeq \infty$, and $-\infty \ll x$ (or $x \gg -\infty$) to mean $x \not\simeq -\infty$. Thus $|x| \ll \infty$ if and only if x is limited, and $|x| \simeq \infty$ if and only if x is unlimited.

Visualize the relations $\simeq$, $\underset{\sim}{<}$, and $\ll$ on the real number line. To the naked eye, $\simeq$ looks like $=$, $\underset{\sim}{<}$ looks like $\leq$, and $\ll$ looks like $<$.

We list below a sequence of propositions that follow easily from the definitions given above:

1. $x \simeq 0$ if and only if x is infinitesimal.

2. $x \simeq 0$ if and only if for all $\varepsilon \gg 0$ we have $|x| \leq \varepsilon$.

3. Infinitesimals are limited.

4. Let $x \neq 0$. Then $x \simeq 0$ if and only if $1/x$ is unlimited.

5. $|x| \simeq \infty$ if and only if $1/x \simeq 0$.

6. If x and y are limited, then so are $x + y$ and xy.

7. If x and y are infinitesimal, then so are $x + y$ and xy.

8. If $x \simeq 0$ and $|y| \ll \infty$, then $xy \simeq 0$.

9. $x \underset{\sim}{<} y$ and $y \underset{\sim}{<} x$ if and only if $x \simeq y$.

10. If $x \simeq y$ and $y \simeq z$, then $x \simeq z$.

11. For all n in $\mathbf{N}$, n is standard if and only if n is limited.

12. For all n in $\mathbf{N}$, n is nonstandard if and only if n is unlimited.

Theorem 5.1 *If n is standard and $x_i \simeq y_i$ for $i = 1, \ldots, n$, then*

$$\sum_{i=1}^{n} x_i \simeq \sum_{i=1}^{n} y_i.$$

Proof. We need to show that $\sum_{i=1}^{n}(x_i - y_i) \simeq 0$. Use (7) and external induction. $\square$

This is not true in general for an unlimited n: take $x_i = 1/n$ and $y_i = 0$.

If $x \neq 0$ and $y \neq 0$, we say that $x \sim y$ (x is *asymptotic* to y) in case $x/y \simeq 1$.

Theorem 5.2 *If $0 \ll |x|, |y| \ll \infty$ (that is, if x and y are non-infinitesimal and limited), then $x \sim y$ if and only if $x \simeq y$.*

Proof. Let $0 \ll |x|, |y| \ll \infty$. Suppose that $x \sim y$. Then $x/y = 1 + \alpha$ with $\alpha \simeq 0$, so that $x - y = \alpha y$. But $\alpha y \simeq 0$ by (8), and so $x \simeq y$.

Conversely, suppose that $x \simeq y$. Then $x - y = \alpha$ with $\alpha \simeq 0$, so that $x/y = 1 + \alpha/y$. But $\alpha/y \simeq 0$ by (5) and (8), and so $x \sim y$.

Theorem 5.3 *If $x_i > 0$, $y_i > 0$, and $x_i \sim y_i$ for $i = 1, \ldots, n$, then*

$$\sum_{i=1}^{n} x_i \sim \sum_{i=1}^{n} y_i.$$

Proof. Let $x > 0$ and $y > 0$. Then $x \sim y$ if and only if for all $\varepsilon \gg 0$ we have $(1 - \varepsilon)y \leq x \leq (1 + \varepsilon)y$. Let $\varepsilon \gg 0$. Then

$$(1 - \varepsilon)y_i \leq x_i \leq (1 + \varepsilon)y_i.$$

Therefore

$$(1 - \varepsilon)\sum_{i=1}^{n} y_i \leq \sum_{i=1}^{n} x_i \leq (1 + \varepsilon)\sum_{i=1}^{n} y_i. \quad \square$$

There is no requirement in Theorem 5.3 that n be limited. Theorems 5.2 and 5.3 together are the reason why integration works: we make only an infinitesimal error in adding up an unlimited number of infinitesimals, provided only that the sum is limited and that we make an infinitesimal *percentage* error in each summand.

I have emphasized that the rules of our theory do not allow us to form subsets corresponding to external properties. In many cases we can even show that such subsets do not exist:

Theorem 5.4 *There does not exist a set A_1, A_2, A_3, A_4, or A_5 such that (for all n and x)*

$n \in A_1$ if and only if $n \in \mathbf{N}$ and n is standard,
$n \in A_2$ if and only if $n \in \mathbf{N}$ and n is nonstandard,
$x \in A_3$ if and only if $x \in \mathbf{R}$ and x is limited,
$x \in A_4$ if and only if $x \in \mathbf{R}$ and x is unlimited,
$x \in A_5$ if and only if $x \in \mathbf{R}$ and x is infinitesimal.

Proof. The existence of A_1 would violate the induction theorem, if A_2 existed we could take $A_1 = \mathbf{N} \setminus A_2$, if A_3 existed we could take $A_2 = \mathbf{N} \setminus A_3$, if A_4 existed we could take $A_3 = \mathbf{R} \setminus A_4$, and if A_5 existed we could take $A_4 = \{x \in \mathbf{R} : 1/x \in A_5\}$. $\square$

This seems like a negative result, but it is very useful. For example, if $A(x)$ is an *internal* formula and we have shown that for all infinitesimal x we have $A(x)$, then we know that there exists a non-infinitesimal x such that $A(x)$, because otherwise we could let $A_5 = \{x \in \mathbf{R} : A(x)\}$. This is called *overspill*.

Let $x_1, x_2, \ldots$ be a sequence of real numbers such that $x_n \simeq 0$ for all limited n. Since "$x_n \simeq 0$" is an external formula, we cannot use overspill directly to prove that $x_n \simeq 0$ for some unlimited n. Nevertheless, the result, called *Robinson's lemma*, is true, and the proof is accomplished by replacing "$x_n \simeq 0$" with a weaker internal formula:

Theorem 5.5 *Let $x_1, x_2, \ldots$ be a sequence of real numbers such that $x_n \simeq 0$ for all limited n. Then there is an unlimited ν such that $x_n \simeq 0$ for all $n \leq \nu$.*

Proof. Consider the set S of all m such that $|x_n| \leq 1/n$ for all $n \leq m$. This set contains all standard numbers and therefore, by overspill (Theorem 5.4), it contains a nonstandard (unlimited) natural number ν. Let $n \leq \nu$. If n is limited, then $x_n \simeq 0$ by hypothesis. If n is unlimited, then $|x_n| \leq 1/n \simeq 0$, since ν is in S. Thus $\nu \simeq \infty$ and $x_n \simeq 0$ for all $n \leq \nu$. $\square$

If we attempt to construct a counterexample by considering the sequence such that $x_n = 0$ whenever n is limited and $x_n = 1$ whenever n is unlimited, we commit illegal set formation. There is no such sequence. A sequence is a function, a function is a set, and we may not use external properties to define sets.

Chapter 6

External analogues of internal notions

Let T be a finite subset of $\mathbf{R}$. Throughout this book, we will use the following notation: the first element of T is a, the last is b, we let $T' = T \setminus \{b\}$, for t in T' its successor is $t + dt$, and for any $\xi: T \to \mathbf{R}$ we write $d\xi(t) = \xi(t + dt) - \xi(t)$. Whenever $\xi: T \to \mathbf{R}$ and we write $\xi(t)$, it is understood that t is in T. If $0 \ll b - a \ll \infty$ and each dt is infinitesimal, we call T a *near interval*.

Most of the concepts of analysis are vacuous when applied to a function whose domain is a finite set of points, so there will be no danger of confusion if we use familiar terminology for external analogues of the concepts. But it is sometimes convenient to insert a modifier, such as "near" or "nearly", before the familiar term when it is used for an elementary external analogue.

Let $T = \{1, \ldots, \nu\}$ where ν is an unlimited natural number. This is an elementary analogue of $\mathbf{N}^+ = \mathbf{N} \setminus \{0\}$. By analogy with the notion of convergence for an infinite sequence, we say that $x_1, \ldots, x_\nu$ is (nearly) *convergent* in case there is a number x such that $x_n \simeq x$ for all unlimited $n \leq \nu$. We also say in this case that $x_1, \ldots, x_\nu$ (nearly) *converges to* x. Notice that if it converges to x, then it also converges to y if and only if $y \simeq x$.

In Fig 6.1 the values of x_n are indistinguishable to the naked eye from x whenever $n \leq \nu$ is unlimited. It may not seem from Fig. 6.1 that near convergence captures the intuitive notion of getting closer and closer to x. What can we say about the values of x_n for $n \ll \infty$? Let $\varepsilon \gg 0$, and let n_ε be the least number such that $|x_n - x| \leq \varepsilon$ for all n with $n_\varepsilon \leq n \leq \nu$. Then $n_\varepsilon \ll \infty$, for if it were true that $n_\varepsilon \simeq \infty$ then we would have $n_\varepsilon - 1 \simeq \infty$ and $|x_{n_\varepsilon - 1} - x| \leq \varepsilon$. See Fig. 6.2.

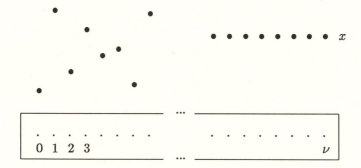

Figure 6.1: Near convergence

There is no unique way to construct an elementary external analogue of a given internal notion. Let T be a subset of $\mathbf{R}$, and let $\xi : T \to \mathbf{R}$. We say that ξ *admits k ε-fluctuations* in case there exist elements $t_0 < \cdots < t_k$ of T with

$$|\xi(t_0) - \xi(t_1)| \geq \varepsilon, \ |\xi(t_1) - \xi(t_2)| \geq \varepsilon, \ \ldots, \ |\xi(t_{k-1}) - \xi(t_k)| \geq \varepsilon$$

(in which case we say that $t_0, \ldots, t_k$ are *indices of k ε-fluctuations*). These are internal notions. Now an infinite sequence is convergent if and only if for all $\varepsilon > 0$ there is a k such that the sequence does not admit k ε-fluctuations. This suggests the following external definition. We say that an infinite or finite sequence (or any function $\xi : T \to \mathbf{R}$) is *of limited fluctuation* in case for all $\varepsilon \gg 0$ and $k \simeq \infty$ it does not admit k ε-fluctuations. Thus the property of a sequence being of limited fluctuation is, like the property of near convergence, an external analogue of the internal notion of convergence. But it is a weaker property: let $i \leq \nu$ be unlimited, and

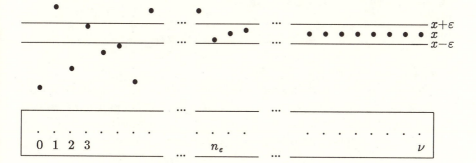

Figure 6.2: Near convergence again

let $x_n = 0$ for $n \leq i$ and $x_n = 1$ for $i < n \leq \nu$. Then $x_1, \ldots, x_\nu$ is of limited fluctuation but it is not convergent. On the other hand, if $x_1, \ldots, x_\nu$ is convergent, then it is easily seen (see Fig. 6.2) to be of limited fluctuation.

Both the notion of near convergence and the notion of being of limited fluctuation will be important for our study of fluctuations in stochastic processes. Near convergence expresses an ordinal property of convergence (from some point on there is no ε-fluctuation), while being of limited fluctuation expresses a cardinal property of convergence (there is only a limited number of ε-fluctuations).

Let $A(n)$ be any formula, internal or external. The *external least number principle* asserts that if there is a standard number n such that $A(n)$, then there is a least number m (which must be standard) such that $A(m)$. This can be proved by external induction, just as the usual least number principle is proved by induction.

*Theorem 6.1 *Let $x_1, \ldots, x_\nu$ be of limited fluctuation, where $\nu \simeq \infty$. Then there is an unlimited $\mu \leq \nu$ such that $x_1, \ldots, x_\mu$ converges.*

Proof. Let $j \ll \infty$. The set of all k such that $x_1, \ldots, x_\nu$ does not admit k $(1/j)$-fluctuations contains all $k \simeq \infty$, so it contains some $k \ll \infty$ by overspill. By the external least number principle, there is a least number l, with $l \leq k$, such that there do not exist *unlimited* indices of $l + 1$ $(1/j)$-fluctuations. If $l = 0$, let $\mu_j = \nu$. If $l > 0$, there are unlimited numbers $n_0, \ldots, n_l$ that are indices of l $(1/j)$-fluctuations; in this case, let $\mu_j = n_0$.

Then $\infty \simeq \mu_j \leq \nu$, and for all unlimited $n \leq \mu_j$ we have $|x_n - x_{\mu_j}| < 1/j$. By the sequence principle, there is a sequence of natural numbers $j \mapsto \mu_j$ such that for all $j \ll \infty$ these properties hold. Now make the sequence decreasing: let $\tilde{\mu}_j = \inf_{l \leq j} \mu_l$. By Robinson's lemma, applied to the reciprocals of the $\tilde{\mu}_j$, there is an unlimited j such that $\tilde{\mu}_j$ is unlimited. Let μ be $\tilde{\mu}_j$ for such a j. Thus μ is unlimited, $\mu \leq \nu$, and $\mu \leq \mu_j$ for all $j \ll \infty$. For all unlimited $n \leq \mu$ we have

$$|x_n - x_\mu| \leq |x_n - x_{\mu_j}| + |x_{\mu_j} - x_\mu| \leq \frac{2}{j}$$

for all $j \ll \infty$, so that $x_n \simeq x_\mu$. Thus $x_1, \ldots, x_\mu$ converges. $\square$

We say that $\sum_{i=1}^{\nu} x_i$ is (nearly) *convergent* in case the sequence of partial sums $y_n = \sum_{i=1}^{n} x_i$, for $n = 1, \ldots, \nu$, is convergent, and is *of limited fluctuation* in case the latter is. (Since "$\sum_{i=1}^{\nu} x_i$" denotes a number and not a sequence, there is an abuse of language here.) Notice that if $\sum_{i=1}^{\nu} x_i$ converges, then it converges to its sum $x = \sum_{i=1}^{\nu} x_i$ (and to any y that is infinitely close to x), and that $\sum_{i=1}^{\nu} x_i$ converges if and only if the tails $\sum_{i=n}^{\nu} x_i$ are infinitesimal for all unlimited $n \leq \nu$.

The following implications are always valid, but none of the reverse implications hold in general.

$$\sum_{i=1}^{\nu} |x_i| \text{ converges} \quad \Rightarrow \quad \sum_{i=1}^{\nu} x_i \text{ converges}$$
$$\Downarrow \qquad\qquad\qquad\qquad \Downarrow$$
$$\sum_{i=1}^{\nu} |x_i| \ll \infty \;\Rightarrow\; \sum_{i=1}^{\nu} |x_i| \;\substack{\text{is of limited}\\\text{fluctuation}} \;\Rightarrow\; \sum_{i=1}^{\nu} x_i \;\substack{\text{is of limited}\\\text{fluctuation}}$$

If $|x_i| \ll \infty$ for all $i \ll \infty$, then we also have the implication

$$\sum_{i=1}^{\nu} |x_i| \text{ converges} \quad \Rightarrow \quad \sum_{i=1}^{\nu} |x_i| \ll \infty.$$

This is because the least n such that $\sum_{i=n}^{\nu} |x_i| \le 1$ must be limited, and

$$\sum_{i=1}^{n-1} |x_i| \le (n-1) \max_{1 \le i \le n-1} |x_i| \ll \infty.$$

Finally, if $|x_i| \ll \infty$ for all $i = 1, \ldots, \nu$, then it is easily seen that

$$\sum_{i=1}^{\nu} |x_i| \ll \infty \quad \Longleftrightarrow \quad \sum_{i=1}^{\nu} |x_i| \text{ is of limited fluctuation.}$$

Let $\xi : T \to \mathbf{R}$. We say that ξ is (nearly) *continuous at* t in case whenever $s \simeq t$ we have $\xi(s) \simeq \xi(t)$. We say that ξ is (nearly) *continuous* (on T) in case it is continuous at each t in T. For example, let T be a near interval with $a > 0$, and let $\xi(t) = 1/t$. Then ξ is continuous at t if and only if $t \gg 0$, so ξ is continuous if and only if $a \gg 0$.

Let ξ be continuous at t and let $\varepsilon \gg 0$. Let Δ_t be the set of all δ such that for all s, if $|s - t| \le \delta$ then $|\xi(s) - \xi(t)| \le \varepsilon$. Then Δ_t contains all $\delta \simeq 0$, so by overspill it contains some $\delta \gg 0$. Thus ξ is continuous at t if and only if for all $\varepsilon \gg 0$ there exists a $\delta \gg 0$ such that for all s, if $|s-t| \le \delta$ then $|\xi(s) - \xi(t)| \le \varepsilon$. Now let ξ be continuous on T, and let $\varepsilon \gg 0$. For each t in T let δ_t be the largest reciprocal of an integer in Δ_t. Then $\delta_t \gg 0$. Let $\delta = \min_t \delta_t$. Then δ is also strongly positive (simply because it is equal to δ_t for some t in the finite set T). Thus ξ is continuous on T if and only if for all $\varepsilon \gg 0$ there is a $\delta \gg 0$ such that for all s and t, $|s - t| \le \delta$ implies $|\xi(s) - \xi(t)| \le \varepsilon$. Hence near continuity at t is an elementary analogue of continuity, and near continuity on T is an elementary analogue of uniform continuity.

Here are some simple and useful illustrations of these notions. By external induction, $e^n \ll \infty$ for all standard n in $\mathbf{N}$. Therefore $e^t \ll \infty$ for all $t \ll \infty$. By the mean value theorem, $e^{t+h} = e^t + e^{\bar{t}} h$ with $\bar{t}$ between t

and $t + h$. Thus if $t \ll \infty$ and $h \simeq 0$, we have $e^{t+h} \simeq e^t$, so that $t \mapsto e^t$ is continuous on T if $b \ll \infty$. Similarly,

$$\log(t + h) = \log t + h/\bar{t} \simeq \log t$$

for $h \simeq 0$ and $0 \ll t$, so that $t \mapsto \log t$ is continuous on T if $a \gg 0$. Let $|x| \ll \infty$ and $n \simeq \infty$. Then

$$\log \left(1 + \frac{x}{n}\right)^n = n \log \left(1 + \frac{x}{n}\right) = n \left(\frac{x}{n} - \frac{1}{2}\left(\frac{\bar{x}}{n}\right)^2\right) \simeq x,$$

where $\bar{x}$ is between 0 and x, so that

$$\left(1 + \frac{x}{n}\right)^n \simeq e^x.$$

Thus $n \mapsto (1 + x/n)^n$, for $n = 1, \dots, \nu$ where $\nu \simeq \infty$ and x is limited, converges to e^x. By the Lagrange form of the remainder for Taylor series,

$$e^x = \sum_{n=0}^{\nu} \frac{x^n}{n!} + e^{\bar{x}} \frac{x^{\nu+1}}{(\nu+1)!}$$

with $\bar{x}$ between 0 and x. If x is limited so is $e^{\bar{x}}$, and if $\nu \simeq \infty$ it is easy to see that the remainder is infinitesimal. Consequently, if $|x| \ll \infty$ and $\nu \simeq \infty$, then $\sum_{n=0}^{\nu} x^n/n!$ converges to e^x.

The *variation*, or *total variation*, of $\xi: T \to \mathbf{R}$ is $\sum_{t \in T'} |d\xi(t)|$. We say that ξ is *of limited variation* in case its variation is limited. This is an elementary analogue of the notion of a function of bounded variation. We say that ξ is (nearly) *absolutely continuous* in case whenever $S \subseteq T'$ is such that $\sum_{t \in S} dt \simeq 0$, then $\sum_{t \in S} |d\xi(t)| \simeq 0$. Clearly, an absolutely continuous function is continuous.

Theorem 6.2 *Let T be a near interval. If ξ is absolutely continuous, then ξ is of limited variation.*

Proof. Consider the set of all δ such that whenever $S \subseteq T'$ with $\sum_{t \in S} dt \leq \delta$, then $\sum_{t \in S} |d\xi(t)| \leq 1$. Since this set contains all $\delta \simeq 0$, it contains some $\delta \gg 0$. Then T' can be be split into $n = [(b-a)/\delta] + 1$ subsets S, with the last element of each S equal to the first element of the next S, with $\sum_{t \in S} |d\xi(t)| \leq 1$. Thus $\sum_{t \in T'} |d\xi(t)| \leq n \ll \infty$. $\square$

Chapter 7

Properties that hold almost everywhere

Finite probability spaces are usually employed only in the discussion of combinatorial problems, but we want to make them the basis for a discussion of the classical limit theorems of probability theory and of central topics in the modern theory of stochastic processes. The fundamental external notion that enables us to do this is the following. Let $\langle \Omega, \mathrm{pr} \rangle$ be a finite probability space and let $A(\omega)$ be an internal or external formula. We say that $A(\omega)$ holds *almost everywhere* (a.e.), or *almost surely* (a.s.), on $\langle \Omega, \mathrm{pr} \rangle$ in case for all $\varepsilon \gg 0$ there is an event N with $\mathrm{Pr}\, N \leq \varepsilon$ such that $A(\omega)$ holds for all ω in N^{c}.

If $A(\omega)$ is an internal formula, then we can form the set

$$\{A\} = \{\omega \in \Omega : A(\omega)\},$$

and $A(\omega)$ holds a.e. if and only if $\mathrm{Pr}\{A\} \simeq 1$. But some of the most interesting properties that we shall consider are external, and we need the formulation of the preceding paragraph to avoid illegal set formation. Whether $A(\omega)$ is internal or external, though, the intuitive content of the statement that $A(\omega)$ holds a.e. is near certainty: given $\varepsilon \gg 0$—for example, $\varepsilon = 10^{-100}$—there is an event N with $\mathrm{Pr}\, N \leq \varepsilon$ such that with the possible exception of points in N the formula $A(\omega)$ always holds.

Theorem 7.1 *Let x be a random variable. Then the following are equivalent*:

    (i) $x \simeq 0$ *a.e.*,

    (ii) *for all $\lambda \gg 0$ we have* $\mathrm{Pr}\{|x| \geq \lambda\} \simeq 0$,

    (iii) *there is a $\lambda \simeq 0$ such that* $\mathrm{Pr}\{|x| \geq \lambda\} \simeq 0$.

Proof. Suppose (i), and let $\lambda \gg 0$ and $\varepsilon \gg 0$. Then there is an event N

with $\Pr N \leq \varepsilon$ such that $x(\omega) \simeq 0$ for all ω in N^c, so that $\{|x| \geq \lambda\} \subseteq N$ and thus $\Pr\{|x| \geq \lambda\} \leq \varepsilon$. Since $\varepsilon \gg 0$ is arbitrary, $\Pr\{|x| \geq \lambda\} \simeq 0$. Thus (i) $\Rightarrow$ (ii). Suppose (ii). Then the set of all λ such that $\Pr\{|x| \geq \lambda\} \leq \lambda$ contains all $\lambda \gg 0$ and so contains some $\lambda \simeq 0$ by overspill. Thus (ii) $\Rightarrow$ (iii). Finally, the implication (iii) $\Rightarrow$ (i) is obvious. $\square$

So long as we are considering a single random variable x, if $x \simeq 0$ a.e. then we can safely think of x as being 0 for all practical purposes—the probability of being able to detect with the naked eye any difference from 0 is less than 10^{-100}. The situation changes radically when we consider an unlimited number of random variables $x_1, \ldots, x_\nu$ each of which is infinitesimal a.e. Suppose that the day is divided into ν equal parts of infinitesimal duration $1/\nu$, that we have a device whose malfunction would cause a disaster, that the probability of malfunction in any period is c/ν where $0 \ll c \ll \infty$, and that different periods are independent. If we let x_n be the indicator function of the event of a malfunction in the n'th period, then for each n we have $x_n \simeq 0$ a.e. (in fact, $x_n = 0$ a.e.). But we are really interested in $\max x_n$, the indicator function of a disaster sometime during the day. By independence, the probability of no disaster during the day is

$$\left(1 - \frac{c}{\nu}\right)^\nu \simeq e^{-c} \ll 1.$$

Let $x_1, \ldots, x_\nu$ be a finite sequence of random variables, with ν unlimited. We say that $x_1, \ldots, x_\nu$ (nearly) *converges to x in probability* in case $x_n \simeq x$ a.e. for all unlimited $n \leq \nu$. As the example above shows, this is not very restrictive. A more interesting question is whether $x_1, \ldots, x_\nu$ converges to x a.e. For convergence in probability the exceptional set N is allowed to depend on n, but not for convergence a.e.

Theorem 7.2 *Let $x_1, \ldots, x_\nu$ be random variables. Then $x_1, \ldots, x_\nu$ converges to 0 a.e. if and only if for all $\lambda \gg 0$ and all unlimited $n \leq \nu$ we have*

$$\Pr\left\{\max_{n \leq i \leq \nu} |x_i| \geq \lambda\right\} \simeq 0. \tag{7.1}$$

Proof. Let

$$M(n, \lambda) = \left\{\max_{n \leq i \leq \nu} |x_i| \geq \lambda\right\}.$$

Suppose that $x_1, \ldots, x_\nu$ converges to 0 a.e., and let $\lambda \gg 0$ and $\varepsilon \gg 0$. Then there is an event N with $\Pr N \leq \varepsilon$ such that $x_1, \ldots, x_\nu$ converges to 0 on N^c. Then $M(n, \lambda) \subseteq N$ if n is unlimited, so that $\Pr M(n, \lambda) \leq \varepsilon$. Since $\varepsilon \gg 0$ is arbitrary, $\Pr M(n, \lambda) \simeq 0$.

Conversely, suppose that $\Pr M(n, \lambda) \simeq 0$ for $n \simeq \infty$ and $\lambda \gg 0$. Let $\varepsilon \gg 0$, and for $j \neq 0$ in $\mathbf{N}$, let n_j be the least natural number such that

$$\Pr M\left(n_j, \frac{1}{j}\right) \leq \frac{\varepsilon}{2^j}.$$

Let

$$N = \bigcup_{j=1}^{\infty} M\left(n_j, \frac{1}{j}\right).$$

(Not that it matters, but the $M(n_j, 1/j)$ are empty for j sufficiently big, since Ω is finite.) Then $\Pr N \leq \varepsilon$. Notice that if j is limited, so is n_j, for otherwise we would have $n_j - 1 \simeq \infty$ and $1/j \gg 0$, so that

$$\Pr M\left(n_j - 1, \frac{1}{j}\right)$$

would be infinitesimal by hypothesis and so $\leq \varepsilon/2^j$, contradicting the definition of n_j. Consequently, if $\omega \in N^c$ and $n \simeq \infty$, then $|x_n(\omega)| \leq 1/j$ for all $j \ll \infty$, and so $x_n(\omega) \simeq 0$. Since $\varepsilon \gg 0$ is arbitrary, this shows that $x_1, \ldots, x_\nu$ converges to 0 a.e. $\square$

Notice that, by Theorem 7.1, the relation (7.1) is equivalent to saying that for all unlimited $n \leq \nu$ we have

$$\max_{n \leq i \leq \nu} |x_i| \simeq 0 \text{ a.e.}$$

Theorem 7.2 has the following corollary:

Corollary. *Let ξ be a stochastic process indexed by a finite subset T of $\mathbf{R}$, and let t be in T. Then ξ is continuous at t a.s. if and only if for all $\lambda \gg 0$ and all $h \simeq 0$ we have*

$$\Pr\left\{\max_{|s-t| \leq h} |\xi(s) - \xi(t)| \geq \lambda\right\} \simeq 0. \tag{7.2}$$

Proof. Let $\nu \simeq \infty$, and for $n \leq \nu$ let

$$x_n = \max_{|s-t| \leq 1/n} |\xi(s) - \xi(t)|.$$

Then ξ is continuous at t if and only if x_n converges to 0, and (7.1) is equivalent to (7.2). $\square$

Theorem 7.3 (Borel-Cantelli, ordinal version) *Let $A_1, \ldots, A_\nu$ be events and let $k(\omega)$ be the largest k such that $\omega \in A_k$, with $k(\omega) = 0$ if ω is not in any of the A_n.*

(i) *If $\sum_{n=1}^{\nu} \Pr A_n$ converges, then k is limited a.s.*

(ii) *If the A_n are independent, then $\sum_{n=1}^{\nu} \Pr A_n$ converges if and only if k is limited a.s.*

Proof. To prove (i), let $\varepsilon \gg 0$ and let j be the least number such that $\Pr \bigcup_{n=j}^{\nu} A_n \leq \varepsilon$. Then j is limited, since otherwise we would have $\Pr \bigcup_{n=j-1}^{\nu} A_n \leq \sum_{n=j-1}^{\nu} \Pr A_n \simeq 0$. But for all ω in $(\bigcup_{n=j}^{\nu} A_n)^c$ we have $k(\omega) \leq j - 1 \ll \infty$. Thus $k \ll \infty$ a.s.

In case (ii) we have

$$\Pr \bigcap_{n=i}^{\nu} A_n^c = \prod_{n=i}^{\nu} (1 - \Pr A_n) \leq e^{-\sum_{n=i}^{\nu} \Pr A_n}$$

since $1 - \lambda \leq e^{-\lambda}$ (a convex function lies above its tangent line). If $k \ll \infty$ a.s., then for $i \simeq \infty$ the left hand side is infinitely close to 1, and so $\sum_{n=i}^{\nu} \Pr A_n \simeq 0$. Together with (i), this proves (ii). $\square$

Theorem 7.4 (Borel-Cantelli, cardinal version) *Let $A_1, \ldots, A_\nu$ be events and let $K(\omega)$ be the number of n such that $\omega \in A_n$.*

(i) *If $\sum_{n=1}^{\nu} \Pr A_n \ll \infty$, then K is limited a.s.*

(ii) *If the A_n are independent, then $\sum_{n=1}^{\nu} \Pr A_n \ll \infty$ if and only if K is limited a.s. In fact, $\sum_{n=1}^{\nu} \Pr A_n \simeq \infty$ if and only if K is unlimited a.s.*

Proof. To prove (i), notice that $\sum_{n=1}^{\nu} \Pr A_n = \mathbf{E}K$, and by the Chebyshev inequality for $p = 1$ we have $\Pr\{K \geq l\} \leq \mathbf{E}K/l$. But if $\mathbf{E}K \ll \infty$, then for all $\varepsilon \gg 0$ there is an $l \ll \infty$ such that $\mathbf{E}K/l \leq \varepsilon$, so that $K < l \ll \infty$ except for an event of probability at most ε, and hence $K \ll \infty$ a.s.

To prove (ii), we need only show (when the A_n are independent) that if $\mathbf{E}K \simeq \infty$, then $K \simeq \infty$ a.s. By the Chebyshev inequality for $p = 2$ we have $\Pr\{|K - \mathbf{E}K| \geq \lambda\} \leq \operatorname{Var} K/\lambda^2$. But by independence,

$$\operatorname{Var} K = \sum \operatorname{Var} \chi_{A_n} \leq \sum \mathbf{E}\chi_{A_n}^2 = \mathbf{E}K.$$

Hence $\Pr\{|K - \mathbf{E}K| \geq \lambda\} \leq \mathbf{E}K/\lambda^2$. Suppose that $\mathbf{E}K \simeq \infty$ and take $\lambda = \frac{1}{2}\mathbf{E}K$. Then $\Pr\{|K - \mathbf{E}K| \geq \frac{1}{2}\mathbf{E}K\} \simeq 0$. Thus except for an event of infinitesimal probability we have $|K - \mathbf{E}K| < \frac{1}{2}\mathbf{E}K$, so that $K \simeq \infty$ a.s. $\square$

If the A_n are independent, then either $K \ll \infty$ a.s. or $K \simeq \infty$ a.s. This is not in general true for the k of Theorem 7.3: let $A_n = \emptyset$ for all $n \leq \nu$ except for $n = i$, where $i \simeq \infty$, and let $\Pr A_i = \frac{1}{2}$. Then the A_n are independent, but $\Pr\{k = 0\} = \frac{1}{2}$ and $\Pr\{k = i\} = \frac{1}{2}$. In this example $\sum \Pr A_n$ is limited but not convergent.

A convergent series of probabilities converges to a limited value, and if $k \ll \infty$ then $K \ll \infty$, so in the ordinal version both the hypothesis and the conclusion of (i) are stronger. The language of conventional mathematics is not well adapted for distinguishing between the ordinal and cardinal versions: if we have an actual infinity $A_1, A_2, \ldots$ of events, then the cardinal statement that only a finite number K of them occur is equivalent to the ordinal statement that from some point k on none of them occur. Yet the distinction is of some practical importance. The discussion above makes it plain that if we want to estimate K, then a bound on $\sum \Pr A_n$ suffices, whereas if we want to estimate k, then we need an estimate showing the smallness of the tails of $\sum \Pr A_n$. For the disaster example discussed earlier in this chapter, it is not true that k is limited a.s., but at least we can take comfort in the fact that almost surely only a limited number of disasters will occur during the day.

Chapter 8

L^1 random variables

If x is a random variable and a is a constant, we define the *truncated* random variable $x^{(a)}$ by
$$x^{(a)} = x\chi_{\{|x|\le a\}}.$$
Thus $x^{(a)}(\omega) = x(\omega)$ if $|x(\omega)| \le a$ and otherwise $x^{(a)}(\omega) = 0$.

We say that a random variable x *is* L^1 in case $\mathbf{E}|x - x^{(a)}| \simeq 0$ for all $a \simeq \infty$. Since
$$\mathbf{E}|x - x^{(a)}| = \sum_{|\lambda| > a} |\lambda| \mathrm{pr}_x(\lambda),$$
it follows that x is L^1 if and only if the sequence
$$n \mapsto \sum_{|\lambda| \le n} |\lambda| \mathrm{pr}_x(\lambda),$$
for $n = 1, \ldots, \nu$ with $\nu \ge \|x\|_\infty$, converges, or as we shall say more briefly, $\sum |\lambda| \mathrm{pr}_x(\lambda)$ (nearly) *converges*. Thus if x is L^1, then $\mathbf{E}|x| \ll \infty$. The converse is not true in general: suppose that $\mathrm{pr}(\omega) \simeq 0$ and let $x = \mathrm{pr}(\omega)^{-1}\chi_{\{\omega\}}$. Then $\mathbf{E}|x| \ll \infty$, but for $a \simeq \infty$ with $a < \mathrm{pr}(\omega)^{-1}$ we have $\mathbf{E}|x - x^{(a)}| = 1 \not\simeq 0$.

Theorem 8.1 (Radon-Nikodym and converse) *A random variable x is L^1 if and only if $\mathbf{E}|x|$ is limited and for all events M with $\mathrm{Pr}\, M \simeq 0$ we have $\mathbf{E}|x|\chi_M \simeq 0$.*

Proof. Suppose that x is L^1 and $\mathrm{Pr}\, M \simeq 0$. Let $a \simeq \infty$ be such that $a \mathrm{Pr}\, M \simeq 0$ (for example, let $a = 1/\sqrt{\mathrm{Pr}\, M}$). Then
$$\mathbf{E}|x|\chi_M \le \mathbf{E}|x^{(a)}|\chi_M + \mathbf{E}|x - x^{(a)}|\chi_M \le a\mathrm{Pr}\, M + \mathbf{E}|x - x^{(a)}| \simeq 0.$$

Conversely, suppose that $\mathbf{E}|x| \ll \infty$ and for all M of infinitesimal probability we have $\mathbf{E}|x|\chi_M \simeq 0$. Let $a \simeq \infty$ and let $M = \{|x| > a\}$. Then (by Chebyshev's inequality for $p = 1$) we have $\Pr M \leq \mathbf{E}|x|/a \simeq 0$, so that $\mathbf{E}|x|\chi_M \simeq 0$; that is, $\mathbf{E}|x - x^{(a)}| \simeq 0$. $\square$

It follows from this criterion that if x and y are L^1, then so is $x + y$; if x is L^1 and $|y| \leq |x|$, then y is L^1; and if x is L^1 and $||y||_\infty \ll \infty$, then yx is L^1.

Theorem 8.2 (Lebesgue) *If x and y are L^1 and $x \simeq y$ a.e., then $\mathbf{E}x \simeq \mathbf{E}y$.*

Proof. Let $z = x - y$. Then $z \simeq 0$ a.e., so (by Theorem 7.1) there is an $\alpha \simeq 0$ such that $\Pr\{|z| > \alpha\} \simeq 0$. But $|z| \leq |z|\chi_{\{|z| > \alpha\}} + \alpha$, and since z is L^1, Theorem 8.1 implies that $\mathbf{E}|z| \simeq 0$. Hence $\mathbf{E}x \simeq \mathbf{E}y$. $\square$

For $1 < p < \infty$ we say that x *is* L^p in case $|x|^p$ is L^1, and we say that x *is* L^∞ in case $||x||_\infty \ll \infty$. If x is L^p and y is $L^{p'}$, where p' is the conjugate exponent to p, then by the inequality

$$|xy| \leq \frac{1}{p}|x|^p + \frac{1}{p'}|y|^{p'}$$

proved in Chapter 1, the product xy is L^1. Also, if $p \gg 1$ and $\mathbf{E}|x|^p \ll \infty$, then x is L^1, since for $a \simeq \infty$ we have

$$\sum_{|\lambda| > a} |\lambda| \mathrm{pr}_x(\lambda) \leq \sum_{|\lambda| > a} \frac{|\lambda|^p}{a^{p-1}} \mathrm{pr}_x(\lambda) \leq \frac{1}{a^{p-1}} \mathbf{E}|x|^p \simeq 0. \ \square$$

Theorem 8.3 *Let x be L^p where $1 \leq p \leq \infty$, and let $\mathcal{A}$ be an algebra of random variables. Then $\mathbf{E}_{\mathcal{A}}x$ is L^p.*

Proof. For $p = \infty$ this is obvious. For $1 \leq p < \infty$ the relativized Jensen inequality implies that $|\mathbf{E}_{\mathcal{A}}x|^p \leq \mathbf{E}_{\mathcal{A}}|x|^p$, so we need only prove the result for $p = 1$. Let x be L^1. Then $\mathbf{E}|\mathbf{E}_{\mathcal{A}}x| \leq \mathbf{E}|x| \ll \infty$. Let $\Pr M \simeq 0$ and let $a = 1/\sqrt{\Pr M}$. Then

$$\mathbf{E}|\mathbf{E}_{\mathcal{A}}x|\chi_M \leq \mathbf{E}|\mathbf{E}_{\mathcal{A}}x^{(a)}|\chi_M + \mathbf{E}|\mathbf{E}_{\mathcal{A}}(x - x^{(a)})|\chi_M$$

$$\leq a \Pr M + \mathbf{E}|x - x^{(a)}| \simeq 0.$$

By Theorem 8.1, $\mathbf{E}_{\mathcal{A}}x$ is L^1. $\square$

Theorem 8.4 *Let x be L^1 and let $\mathcal{A}$ be an algebra of random variables. Then x is L^1 on a.e. atom of $\mathcal{A}$.*

Proof. Let $\varepsilon \gg 0$. For each n in $\mathbf{N}$ let a_n be the least natural number such that

$$\Pr_{\mathcal{A}}' \left\{ \mathbf{E}_{\mathcal{A}}|x - x^{(a_n)}| \geq \frac{1}{n} \right\} \leq \frac{\varepsilon}{2^n}.$$

(See Chapter 2 for the definition of Pr'_A.) I claim that if $n \ll \infty$ then $a_n \ll \infty$. To see this, observe that $\mathbf{E}_A|x - x^{(a)}|$ is a random variable on $\mathrm{at}(A)$ whose expectation is

$$\mathbf{E}'_A\mathbf{E}_A|x - x^{(a)}| = \mathbf{E}|x - x^{(a)}|.$$

The Chebyshev inequality implies that

$$\mathrm{Pr}'_A\left\{\mathbf{E}_A|x - x^{(a)}| \geq \frac{1}{n}\right\} \leq n\mathbf{E}|x - x^{(a)}|,$$

which is infinitesimal (and so $\leq \varepsilon/2^n$) if $a \simeq \infty$.

Let N be the set of all atoms on which

$$\mathbf{E}_A|x - x^{(a_n)}| \geq \frac{1}{n}$$

for some n. Then $\mathrm{Pr}'_A N \leq \varepsilon$. On those atoms not in N, we have

$$\mathbf{E}_A|x - x^{(a)}| < \frac{1}{n}$$

for all $n \ll \infty$, if $a \simeq \infty$, since $a \geq a_n$ for all $n \ll \infty$. Thus x is L^1 on those atoms. Since $\varepsilon \gg 0$ is arbitrary, this concludes the proof. $\square$

In the converse direction, suppose that x is L^1 on *every* atom. If $a \simeq \infty$, then $\mathbf{E}_A|x - x^{(a)}| \simeq 0$ everywhere, so that $\mathbf{E}|x - x^{(a)}| \simeq 0$. Thus if x is L^1 on every atom of A, then x is L^1. This is the most that can be said in general, since we can always alter an L^1 random variable on a single point of infinitesimal probability and obtain a random variable that is not L^1.

Theorem 8.4 has the following corollary.

Corollary. (Fubini) *If x is L^1 on $\langle\Omega_1 \times \Omega_2, \mathrm{pr}_1 \times \mathrm{pr}_2\rangle$, then the random variable x_{ω_1} on $\langle\Omega_2, \mathrm{pr}_2\rangle$, given by $x_{\omega_1}(\omega_2) = x(\omega_1, \omega_2)$, is L^1 for a.e. ω_1 in the space Ω_1.*

Chapter 9

The decomposition of a stochastic process

We will study stochastic processes ξ indexed by a finite subset T of $\mathbf{R}$. Recall the general notation introduced at the beginning of Chapter 6. Typical cases will be $T = \{1, \ldots, \nu\}$ where ν is an unlimited natural number or the case that T is a near interval. Thus, although we require T to be finite, we will be studying the classical subjects of "infinite" sequences of random variables and of "continuous" time parameter stochastic processes.

Let $P: t \mapsto P_t$ be an increasing function from T to the set of all algebras of random variables on $\langle \Omega, \mathrm{pr} \rangle$. This is called a *filtration*. We abbreviate $\mathbf{E}_{P_t}$ by $\mathbf{E}_t$. A P-*process*, or a *process adapted to* P, is a stochastic process ξ indexed by T such that for all t in T we have $\xi(t) \in P_t$. Since $P_s \subseteq P_t$ for $s \leq t$, if ξ is a P-process, then $\xi(s) \in P_t$ for all $s \leq t$. If ξ is any stochastic process indexed by T, then it is a P-process if we define each P_t to be the algebra generated by the $\xi(s)$ with $s \leq t$, but it is convenient to allow for the possibility that P_t is larger. The algebra P_t represents the past at time t, and if y is a random variable, then $\mathbf{E}_t y$ is the best prediction of y that can be made knowing the past at time t.

A P-process ξ is called a *martingale* in case $\mathbf{E}_s \xi(t) = \xi(s)$ for all $s \leq t$, a *submartingale* in case $\mathbf{E}_s \xi(t) \geq \xi(s)$ for all $s \leq t$, and a *supermartingale* in case $\mathbf{E}_s \xi(t) \leq \xi(s)$ for all $s \leq t$. Thus in the trivial case that Ω consists of a single point, a martingale reduces to a constant function of t, a submartingale to an increasing function, and a supermartingale to a decreasing function.

Notice that if ξ is a martingale, then

$$\xi(t) = \mathbf{E}_t \xi(b)$$

for all t. Conversely, given a filtration $\mathcal{P}$ and any random variable x, the process ξ defined by $\xi(t) = \mathbf{E}_t x$ is a martingale.

Let ξ be a $\mathcal{P}$-process. We define $D\xi$, $d\hat{\xi}$, and σ_ξ^2 by

$$D\xi(t)dt = \mathbf{E}_t d\xi(t),$$

$$d\xi(t) = D\xi(t)dt + d\hat{\xi}(t),$$

$$\sigma_\xi^2(t)dt = \mathbf{E}_t d\hat{\xi}(t)^2.$$

Thus $D\xi(t)dt$ and $\sigma_\xi^2(t)dt$ are the conditional mean and variance of the increment $d\xi(t)$. Notice that $D\xi$ and σ_ξ^2 are $\mathcal{P}$-processes indexed by T', whereas $d\xi$ and $d\hat{\xi}$ are not in general $\mathcal{P}$-processes.

Observe that $D\xi = 0$ if ξ is a martingale, $D\xi \geq 0$ if ξ is a submartingale, and $D\xi \leq 0$ if ξ is a supermartingale. (Of course, for a general process $D\xi$ need not have a constant sign either in t or in ω.) We will show that these conditions are sufficient as well as necessary. We have

$$\xi(t) = \xi(s) + \sum_{s \leq r < t} D\xi(r)dr + \sum_{s \leq r < t} d\hat{\xi}(r), \quad s \leq t. \tag{9.1}$$

Now $\mathbf{E}_r d\hat{\xi}(r) = 0$. Since $\mathcal{P}_s \subseteq \mathcal{P}_r$ for $s \leq r$, we have $\mathbf{E}_s = \mathbf{E}_s \mathbf{E}_r$ for $s \leq r$. Hence $\mathbf{E}_s d\hat{\xi}(r) = 0$ for $s \leq r$. Therefore, if we apply $\mathbf{E}_s$ to (9.1) we obtain

$$\mathbf{E}_s \xi(t) = \xi(s) + \mathbf{E}_s \sum_{s \leq r < t} D\xi(r)dr, \quad s \leq t. \tag{9.2}$$

Therefore ξ is a martingale if and only if $D\xi = 0$, a submartingale if and only if $D\xi \geq 0$, and a supermartingale if and only if $D\xi \leq 0$.

We call $D\xi$ the *trend* of ξ, and if $d\hat{\xi} = 0$ we say that ξ is a *predictable process*. Thus ξ is a predictable process if and only if $\sigma_\xi^2 = 0$ or, equivalently, $d\xi$ is a $\mathcal{P}$-process. We let

$$\tilde{\xi}(t) = \sum_{s < t} D\xi(s)ds,$$

so that $\tilde{\xi}(a) = 0$. Then $d\tilde{\xi}(t) = D\xi(t)dt$ is in $\mathcal{P}_t$, so that $\tilde{\xi}$ is a predictable process. We call it the *predictable process associated with* ξ. Notice that if we know $\mathcal{P}_t$, then we know $d\tilde{\xi}(t)$; we can predict the increment with certainty. But to predict the next increment with certainty, we would need to know $\mathcal{P}_{t+dt}$, and this is not in general generated by $\mathcal{P}_t$ and $d\tilde{\xi}(t)$. We let

$$\hat{\xi}(t) = \xi(a) + \sum_{s < t} d\hat{\xi}(s):$$

Thus $\hat{\xi}$ is a P-process whose increments, as the notation requires, are the $d\hat{\xi}(t)$. Since $D\hat{\xi} = 0$, the process $\hat{\xi}$ is a martingale. We call it the *martingale associated with* ξ. Notice that if ξ is already a martingale, then $\hat{\xi} = \xi$. We have the decomposition

$$\xi = \tilde{\xi} + \hat{\xi}$$

of an arbitrary P-process ξ into a predictable process $\tilde{\xi}$ and a martingale $\hat{\xi}$, and this decomposition is unique if we impose the normalization $\tilde{\xi}(a) = 0$.

If the $d\xi(t)$ are independent, and if P_t is the algebra generated by the $\xi(s)$ with $s \leq t$, then $D\xi(t) = \mathbf{E}_t d\xi(t) = \mathbf{E} d\xi(t)$. Therefore the partial sums $\xi(t) = \sum_{s<t} d\xi(s)$ of independent random variables $d\xi(s)$ of mean 0 form a martingale. To specify such a process up to equivalence, it is only necessary to give the probability distribution of the increments. Here are two examples. In the first example, which we call the *Wiener walk*, we set

$$d\xi(t) = \begin{cases} \sqrt{dt} & \text{with probability } \frac{1}{2}, \\ -\sqrt{dt} & \text{with probability } \frac{1}{2}, \end{cases}$$

and in the second example, which we call the *Poisson walk*, we assume that $dt \leq 1$ for all t and set

$$d\xi(t) = \begin{cases} 1 & \text{with probability } \frac{1}{2}dt, \\ 0 & \text{with probability } 1 - dt, \\ -1 & \text{with probability } \frac{1}{2}dt. \end{cases}$$

Let ξ be a martingale. If $r_1 < r_2$, then $d\xi(r_1) \in P_{r_2}$. Since $\mathbf{E}_{r_2}$ is P_{r_2}-linear, we have $\mathbf{E}_{r_2} d\xi(r_1) d\xi(r_2) = d\xi(r_1) \mathbf{E}_{r_2} d\xi(r_2) = 0$. But $\mathbf{E}_s \mathbf{E}_{r_2} = \mathbf{E}_s$ for $s \leq r_2$, and consequently

$$\mathbf{E}_s d\xi(r_1) d\xi(r_2) = 0, \quad r_1 \neq r_2, \quad s \leq \max\{r_1, r_2\}.$$

Since $\xi(t) - \xi(s) = \sum_{s \leq r < t} d\xi(r)$ for $s \leq t$, this implies that

$$\mathbf{E}_s \big(\xi(t) - \xi(s)\big)^2 = \mathbf{E}_s \sum_{s \leq r < t} \sigma_\xi^2(r) dr, \quad s \leq t.$$

If we take absolute expectations, we see that ξ has orthogonal increments, and

$$\|\xi(t) - \xi(s)\|_2^2 = \sum_{s \leq r < t} \|d\xi(r)\|_2^2, \quad s \leq t.$$

Now let η be any P-process, and consider the P-process ς given by

$$\varsigma(t) = \sum_{s<t} \eta(s) d\xi(s).$$

Then $\varsigma(a) = 0$ and $d\varsigma(t) = \eta(t)d\xi(t)$, so that $D\varsigma(t) = \eta(t)D\xi(t) = 0$. Thus ς is also a martingale, and $\sigma_\varsigma^2(t) = \eta(t)^2\sigma_\xi^2(t)$. We can think of ξ as describing a fair game. At any time t the expected win on the next bet is $\mathbf{E}_t d\xi(t) = 0$. Then the process η represents a gambling strategy. At any time t the gambler decides, on the basis of her knowledge of the past, to multiply her bet by the factor $\eta(t)$. Then $\varsigma(t)$ represents her winnings at time t. No matter how cleverly the gambler chooses her strategy, any possible winning at time t will be offset by a possible loss at time t, since $\mathbf{E}\varsigma(t) = 0$. This fact is the inspiration for the poem "Owed to a Martingale".

A random variable x is called *normalized* in case $\mathbf{E}x = 0$ and $\operatorname{Var} x = 1$. If x is not a constant random variable, then $x = \mu + \sigma\breve{x}$, where μ is the mean of x, σ is the standard deviation of x, and $\breve{x} = (x - \mu)/\sigma$ is a normalized random variable. Such a representation is possible even if x is a constant random variable (that is, even if $\sigma = 0$) by taking for $\breve{x}$ any normalized random variable. There is one exceptional case: if Ω consists of a single point, then it does not admit any normalized random variable, and it must be replaced by a larger finite probability space.

These trivialities have been recounted to serve as a guide in thinking about the more interesting case of a stochastic process ξ. A stochastic process ξ is called *normalized* in case $D\xi = 0$ and $\sigma_\xi^2 = 1$. In particular, if ξ is normalized then it is a martingale. Both the Wiener walk and the Poisson walk are normalized martingales. A stochastic process ξ is called *non-degenerate* in case $\sigma_\xi^2 > 0$ on $T' \times \Omega$. If ξ is a non-degenerate stochastic process, define $\breve{\xi}$ by

$$\breve{\xi}(t) = \sum_{s<t} \sigma_\xi^{-1}(s)d\hat{\xi}(s),$$

where $\sigma_\xi^{-1} = 1/\sqrt{\sigma_\xi^2}$ and $\hat{\xi}$ is the associated martingale. Then $\breve{\xi}$ is normalized, and we have the representation

$$\xi(t) = \xi(a) + \sum_{s<t} D\xi(s)ds + \sum_{s<t} \sigma_\xi(s)d\breve{\xi}(s). \tag{9.3}$$

Even if ξ is degenerate we can achieve such a representation by passing to a larger finite probability space. Let η be any normalized martingale indexed by T and defined over a finite probability space $\langle \Omega', \mathrm{pr}' \rangle$. Define $\sigma_\xi^{-1}(s)$ to be $1/\sqrt{\sigma_\xi^2(s)}$ where $\sigma_\xi^2(s) > 0$, and 0 otherwise, and let $\breve{\xi}$ be the stochastic process defined on $\langle \Omega \times \Omega', \mathrm{pr} \times \mathrm{pr}' \rangle$ by $\breve{\xi}(a) = 0$ and

$$d\breve{\xi}(s) = \sigma_\xi^{-1}(s)d\hat{\xi}(s) + \chi_{\{\sigma_\xi^2(s)=0\}}d\eta(s).$$

Then $\breve{\xi}$ is normalized and we have the representation (9.3).

Chapter 10

The total variation of a process

The discussion in the last chapter was entirely internal. We shall be interested mainly in processes for which $D\xi$ and σ_ξ^2 are limited in suitable norms. Then the increments $d\tilde{\xi}(t) = D\xi(t)dt$ of the associated predictable process will be of order dt while the increments $d\hat{\xi}(t)$ of the associated martingale will be of order $\sqrt{dt}$. The interesting local fluctuations of a stochastic process are in the associated martingale. Nevertheless, in this chapter we begin the study of fluctuations of stochastic processes by estimating $\sum |d\xi(t)|$, a method that is too crude to be of interest for martingales. We will see that for a process ξ whose difference quotient $d\xi/dt$ is L^1 in both ω and t (in a sense that will be made precise), ξ is absolutely continuous a.s. (i.e., almost surely the sample paths are absolutely continuous) and that almost surely the associated martingale remains infinitely close to its initial value. This last property also holds for an increasing process with infinitesimal increments.

Notice that the definition

$$\text{pr}_T(t) = \frac{dt}{b-a}$$

makes $\langle T', \text{pr}_T \rangle$ into a finite probability space. We denote the expectation on this space by $\mathbf{E}_T$. If ξ is a fixed function on T, then ξ is of limited variation if and only if $\mathbf{E}_T |d\xi/dt| \ll \infty$ and, by Theorem 8.1, ξ is absolutely continuous if and only if $d\xi/dt$ is L^1 on $\langle T', \text{pr}_T \rangle$.

Now let ξ be a stochastic process indexed by T, adapted to the filtration $\mathcal{P}$. When $T = \{1, \ldots, \nu\}$, we sometimes use the notation ξ_n for $\xi(n)$.

Theorem 10.1 (i) *If* $\sum \|d\xi(t)\|_1 \ll \infty$, *then* $\sum |d\xi(t)| \ll \infty$ *a.s. (that is, ξ is of limited variation almost surely).*

   (ii) *If* $T = \{1, \ldots, \nu\}$ *and* $\sum \|d\xi_n\|$ *converges, then* ξ_n *converges a.s.*

(iii) *If $d\xi/dt$ is L^1 on $T' \times \Omega$, then ξ is absolutely continuous a.s.*

Proof. In case (i), let $c = \sum \|d\xi(t)\|_1$ and let $x = \sum |d\xi(t)|$, so that $c = \mathbf{E}\, x$. Then (i) asserts that if $\mathbf{E}\,|x| \ll \infty$, then $|x| \ll \infty$ a.s., which is true in general by the Chebyshev inequality.

In case (ii), by Theorem 7.2 we need only show that for all $\lambda \gg 0$ and unlimited $n < \nu$, we have

$$\Pr\{\sum_{i=n}^{\nu-1} |d\xi_i| \ge \lambda\} \simeq 0.$$

But the left hand side is $\le \sum_{i=n}^{\nu-1} \|d\xi_i\|_1/\lambda \simeq 0$.

Case (iii) follows at once from the Fubini theorem (corollary to Theorem 8.4). $\square$

Case (iii) was stated for a general stochastic process, but we shall see (in Theorem 10.4) that the hypothesis is quite restrictive, and implies that ξ is nearly a predictable process, in the sense that for the associated martingale, almost surely $\hat{\xi}(t) \simeq \hat{\xi}(a)$ for all t.

Theorem 10.2 *Let ξ be a stochastic process such that $d\xi/dt$ is L^1 on $T' \times \Omega$. Then $D\xi$ is L^1 on $T' \times \Omega$, and consequently $d\hat{\xi}/dt = d\xi/dt - D\xi$ is L^1 on $T' \times \Omega$.*

Proof. Let P' be the algebra of all random variables y on $T' \times \Omega$ such that for every t we have $y(t, \cdot) \in P_t$. Then

$$D\xi = \mathbf{E}_{P'} \frac{d\xi}{dt},$$

so that the result follows by Theorem 8.3. $\square$

The next result is a truncation lemma. It asserts that under suitable hypotheses we can modify a process slightly (meaning that almost surely the trajectories stay infinitely close for all times) so that the modified process has infinitesimal increments.

Theorem 10.3 *Let ξ be a martingale indexed by a near interval T such that $d\xi/dt$ is L^1 on $T' \times \Omega$. For $\alpha > 0$, let $\xi_{(\alpha)}$ be the martingale with $\xi_{(\alpha)}(a) = \xi(a)$ and*

$$d\xi_{(\alpha)}(t) = d\xi(t)^{(\alpha)} - \mathbf{E}_t d\xi(t)^{(\alpha)},$$

so that $|d\xi_{(\alpha)}(t)| \le 2\alpha$ for all t and all ω. Then there is an infinitesimal α such that a.s. for all t we have $\xi_{(\alpha)}(t) \simeq \xi(t)$.

Proof. By Theorem 10.1 (iii), a.s. ξ is absolutely continuous, and hence continuous. Since T is a near interval, this implies that $\max_t |d\xi(t)| \simeq 0$

a.s. By Theorem 7.1, for any sufficiently large infinitesimal α we have

$$\Pr\{\max_t |d\xi(t)| \geq \alpha\} \simeq 0.$$

Then a.s. we have

$$\xi_{(\alpha)}(t) = \xi(a) + \sum_{s<t} d\xi(s)^{(\alpha)}.$$

Therefore, by the definition of $\xi_{(\alpha)}$, to show that a.s. $\xi_{(\alpha)}(t) \simeq \xi(t)$ for all t, it is enough to show that $\sum |\mathbf{E}_t d\xi(t)^{(\alpha)}| \simeq 0$ a.s., and for this it suffices to show that its expectation is infinitesimal. But

$$\mathbf{E}\sum |\mathbf{E}_t d\xi(t)^{(\alpha)}| = \mathbf{E}\sum |\mathbf{E}_t(d\xi(t) - d\xi(t)^{(\alpha)})|$$

(since ξ is a martingale, so that $\mathbf{E}_t d\xi(t) = 0$), and this is

$$\leq \mathbf{E}\sum \mathbf{E}_t |d\xi(t) - d\xi(t)^{(\alpha)}| = \mathbf{E}\sum \left| \frac{d\xi}{dt} - \left(\frac{d\xi}{dt}\right)^{(\alpha/dt)} \right| dt$$

$$\leq \mathbf{E}\sum \left| \frac{d\xi}{dt} - \left(\frac{d\xi}{dt}\right)^{(c)} \right| dt$$

where $c = \alpha/\max dt$. If $\alpha \gg 0$, then $c \simeq \infty$; and since $d\xi/dt$ is L^1 on $T' \times \Omega$, the expectation is infinitesimal. Therefore the set of all α such that

$$\mathbf{E}\sum |\mathbf{E}_t d\xi(t)^{(\alpha)}| \leq \alpha$$

contains all $\alpha \gg 0$, and so contains all sufficiently large infinitesimal α, by overspill. □

In the proofs of the next two theorems we shall appeal to a result (Theorem 11.1) that will be proved later.

Theorem 10.4 *Let ξ be a stochastic process indexed by a near interval T such that $d\xi/dt$ is L^1 on $T' \times \Omega$, and let $\hat\xi$ be the associated martingale. Then a.s. $\hat\xi(t) \simeq \hat\xi(a)$ for all t.*

Proof. There is no loss of generality in assuming that $\xi(a) = 0$. By Theorems 10.2 and 10.3, we may assume that $\xi = \hat\xi$, and so is a martingale, and is such that for some infinitesimal α we have $|d\xi(t)| \leq \alpha$ for all t and ω. Then

$$\|d\xi(t)\|_2^2 = \mathbf{E}d\xi(t)^2 \leq \alpha \mathbf{E}|d\xi(t)|,$$

and since $d\xi/dt$ is L^1 on $T' \times \Omega$ we have

$$\sum \mathbf{E}|d\xi(t)| = \mathbf{E}\sum \left|\frac{d\xi}{dt}\right| dt \ll \infty.$$

Therefore

$$\|\xi(b)\|_1^2 \le \|\xi(b)\|_2^2 = \sum \|d\xi(t)\|_2^2 \le \alpha \sum \mathbf{E}|d\xi(t)| \simeq 0.$$

By Theorems 7.1 and 11.1 we have $\max_t |\xi(t)| \simeq 0$ a.s. $\square$

Theorem 10.5 *Let ξ be an increasing stochastic process all of whose increments are everywhere infinitesimal, and suppose that $\mathbf{E}(\xi(b) - \xi(a)) \ll \infty$. Let $\hat{\xi}$ be the associated martingale. Then a.s. $\hat{\xi}(t) \simeq \hat{\xi}(a)$ for all t.*

Proof. Again, there is no loss of generality in assuming that $\xi(a) = 0$. Let $\alpha = \max_{t,\omega} |d\xi(t)|$, so that by hypothesis α is infinitesimal. Then

$$\|\hat{\xi}(b)\|_1^2 \le \|\hat{\xi}(b)\|_2^2 = \sum \|d\hat{\xi}(t)\|_2^2 \le \sum \|d\xi(t)\|_2^2$$

$$= \sum \mathbf{E}d\xi(t)^2 \le \alpha \sum \mathbf{E}d\xi(t) = \alpha\mathbf{E}(\xi(b) - \xi(a)) \simeq 0.$$

By Theorems 7.1 and 11.1 we have $\max_t |\hat{\xi}(t)| \simeq 0$ a.s. $\square$

Chapter 11

Convergence of martingales

In this chapter we derive an estimate on the maximum of a supermartingale or submartingale, and use it to study convergence and continuity properties at a fixed time.

The price on the stock market of one share of Nonstandard Oil is a stochastic process ξ. A bullish investor buys a share at time a and decides to keep it until the price increases by at least λ dollars, at which time he takes his profit and sells it. Then the investor's "earnings" at time t are given by

$$\varsigma_1(t) = \sum_{s<t} \eta_1(s) d\xi(s),$$

where $\eta_1(s)$ is 1 as long as he holds on to the stock and is 0 thereafter. Whether $\eta_1(s)$ is 0 or 1 depends only on the values of $\xi(r)$ for $r \leq s$, so η_1 is a P-process. Recall the general discussion of sums of this form ("stochastic integrals") in Chapter 9. Let Λ be the event that the investor's strategy is successful:

$$\Lambda = \{\max(\xi(t) - \xi(a)) \geq \lambda\}.$$

The investor's final earnings are at least λ if he is successful, and they are $\xi(b) - \xi(a)$ otherwise. That is,

$$\varsigma_1(b) \geq \lambda \chi_\Lambda + (\xi(b) - \xi(a))\chi_{\Lambda^c}. \tag{11.1}$$

Now suppose that, unfortunately for the investor, ξ is a supermartingale (declining market). Since η_1 is a positive P-process, we have

$$D\varsigma_1(t) = \eta_1(t) D\xi(t) \leq 0,$$

so that ς_1 is also a supermartingale. Since $\varsigma_1(a) = 0$, we have $\mathbf{E}\varsigma_1(b) \leq 0$. By (11.1), this implies that if ξ is a supermartingale and $\lambda > 0$, then

$$\Pr\{\max(\xi(t) - \xi(a)) \geq \lambda\} \leq \frac{1}{\lambda}\|\xi(b) - \xi(a)\|_1. \tag{11.2}$$

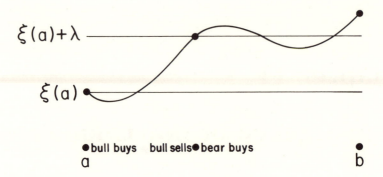

$\xi(a)+\lambda$

$\xi(a)$

● bull buys   bull sells ● bear buys   ●
a   b

Figure 11.1: Two investment strategies

The argument appears to yield no information if ξ is a submartingale. But consider a bearish investor who adopts the opposite strategy: she cautiously waits to see whether the price will increase by at least λ dollars, and as soon as this happens she buys a share and holds on to it. Her earnings at time t are given by

$$\varsigma_2(t) = \sum_{s<t} \eta_2(s)d\xi(s),$$

where $\eta_2(s) = 1 - \eta_1(s)$, so that η_2 is also a positive P-process. Her final earnings are 0 unless the stock rises (the event Λ occurs), in which case they are at most $\xi(b) - \xi(a) - \lambda$. That is,

$$\varsigma_2(b) \leq (\xi(b) - \xi(a) - \lambda)\chi_\Lambda. \qquad (11.3)$$

Notice that (11.3) is the same as (11.1), since $\varsigma_1 + \varsigma_2 = \xi - \xi(a)$. Now if ξ is a submartingale, then so is ς_2, since $D\varsigma_2(t) = \eta_2(t)D\xi(t) \geq 0$. Since $\varsigma_2(a) = 0$, we have $\mathbf{E}\varsigma_2(b) \geq 0$. By (11.3), therefore, if ξ is a submartingale and $\lambda > 0$, then (11.2) holds.

Thus (11.2) holds for both supermartingales and for submartingales (and also, of course, for martingales, either because a martingale is a supermartingale or because a martingale is a submartingale). Since $-\xi$ is a supermartingale if and only if ξ is a submartingale, and vice versa, we also have the inequality

$$\Pr\{\max(\xi(a) - \xi(t)) \geq \lambda\} \leq \frac{1}{\lambda}\|\xi(b) - \xi(a)\|_1,$$

valid for both submartingales and supermartingales. We have proved the following theorem.

Theorem 11.1 *Let ξ be a supermartingale or a submartingale, and let $\lambda > 0$. Then*

$$\Pr\{\max(\xi(t) - \xi(a)) \geq \lambda\} \leq \frac{1}{\lambda}\|\xi(b) - \xi(a)\|_1,$$

$$\Pr\{\max|\xi(t) - \xi(a)| \geq \lambda\} \leq \frac{2}{\lambda}\|\xi(b) - \xi(a)\|_1. \qquad (11.4)$$

If ξ is a martingale, then, by the following theorem, we can prove a result that is twice as good as (11.4).

Theorem 11.2 (i) *If ξ is a martingale and f is a convex function, then $f \circ \xi$ is a submartingale.*

(ii) *If ξ is a submartingale and f is a convex increasing function, then $f \circ \xi$ is a submartingale.*

Proof. By the relativization of Jensen's inequality, if f is convex, then $f(\mathbf{E}_s\xi(t)) \leq \mathbf{E}_s f(\xi(t))$. If ξ is a martingale, then $\mathbf{E}_s\xi(t) = \xi(s)$ for $s \leq t$, so that $f(\xi(s)) \leq \mathbf{E}_s f(\xi(t))$ for $s \leq t$, and $f \circ \xi$ is a submartingale. If ξ is a submartingale, then $\xi(s) \leq \mathbf{E}_s\xi(t)$ for $s \leq t$, and if f is increasing and convex, then $f(\xi(s)) \leq f(\mathbf{E}_s\xi(t)) \leq \mathbf{E}_s f(\xi(t))$ for $s \leq t$ (by the relativized Jensen inequality again), and $f \circ \xi$ is a submartingale. $\square$

If ξ is a martingale, then $\xi - \xi(a)$ is a martingale, $|\xi - \xi(a)|$ is a submartingale by (ii), and so by (11.2) the inequality (11.4) holds with $2/\lambda$ replaced by $1/\lambda$.

The set $\mathbf{R}^\Omega$ of all random variables on $\langle\Omega, \mathrm{pr}\rangle$ is a metric space with respect to the metric $\|x - y\|_p$, where $1 \leq p \leq \infty$. We denote this metric space by $\mathbf{L}^p$. As p varies, the $\mathbf{L}^p$ are the same set $\mathbf{R}^\Omega$, but with different metrics. We say that $x_1, \ldots, x_\nu$ (nearly) *converges to y in $\mathbf{L}^p$* in case we have $\|x_n - y\|_p \simeq 0$ for all unlimited $n \leq \nu$. If ξ is a stochastic process indexed by the finite subset T of $\mathbf{R}$, and $t \in T$, we say that ξ is (nearly) *continuous at t in $\mathbf{L}^p$* in case $s \simeq t$ implies $\|\xi(s) - \xi(t)\|_p \simeq 0$, and otherwise we say that ξ is (strongly) *discontinuous at t in $\mathbf{L}^p$*.

Continuity in $\mathbf{L}^p$ is an analytical property that is easy to verify or falsify, but almost sure continuity of sample paths is an interesting and subtle probabilistic property. Many theorems of the theory of stochastic processes assert that under suitable analytical hypotheses, a certain property of sample paths holds almost surely.

Theorem 11.3 (i) *Let $x_1, \ldots, x_\nu$ be a supermartingale or submartingale that converges in $\mathbf{L}^1$. Then it converges a.s.*

(ii) *Let ξ be a supermartingale or submartingale that is continuous at t in $\mathbf{L}^1$. Then it is continuous at t a.s.*

Proof. This follows immediately from Theorem 11.1 and from Theorem 7.2 and its corollary. $\square$

The next theorem establishes a partial converse.

Theorem 11.4 (i) *Let $x_1, \ldots, x_\nu$ be a martingale such that x_ν is L^1. Then $x_1, \ldots, x_\nu$ converges in $\mathbf{L}^1$ if and only if it converges a.s.*

(ii) *Let ξ be a martingale such that $\xi(b)$ is L^1. Then ξ is continuous in $\mathbf{L}^1$ at t if and only if ξ is continuous a.s. at t.*

Proof. By Theorem 8.3, $\xi(t) = \mathbf{E}_t \xi(b)$ is L^1 for all t. By the Lebesgue theorem (Theorem 8.2), a.s. convergence or continuity then implies $\mathbf{L}^1$ convergence or continuity. The other direction was proved in Theorem 11.3. $\square$

The only use made in this proof of the martingale hypothesis was to conclude that every $\xi(t)$ is L^1, so the result continues to hold with this as the hypothesis.

We say that t is a (strong) *fixed point of discontinuity* of ξ in case it is not true that ξ is continuous a.s. at t. Thus Theorem 11.4 (ii) says that if ξ is a martingale such that $\xi(b)$ is L^1, then t is a fixed point of discontinuity of ξ if and only if t is a discontinuity of ξ in $\mathbf{L}^1$. The following counterexample shows what can happen when $\xi(b)$ is not L^1.

Let Ω be an ordered set of 2^ν points, where $\nu \simeq \infty$, each with the

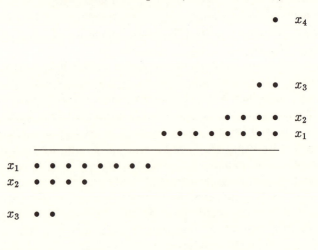

Figure 11.2: A bad martingale

probability $2^{-\nu}$, and let x_n, for $n = 1, \ldots, \nu$, be the random variable that takes the value -2^{n-1} on the first $2^{\nu-n}$ points, the value 2^{n-1} on the last $2^{\nu-n}$ points, and the value 0 in between. (The non-zero values are illustrated in Fig. 11.2, where to save paper we have taken $\nu = 4$ rather that $\nu \simeq \infty$.) Then the x_n are a martingale, with respect to the algebras P_n generated by $x_1, \ldots, x_n$. They have mean 0 and $\|x_n\|_1 = 1$ for all n, and $\|x_n - x_m\|_1 \geq 1$ for all $n \neq m$. Let k and $\nu - k$ be unlimited, and let T be a near interval containing $\nu - k$ points. Define ξ by $\xi(t) = x_{n+k}$, where t is the n'th point of T. Then ξ is everywhere discontinuous in $\mathbf{L}^1$, but since a.s. $\xi(t) = 0$ for all t, it is a.s. continuous for all t.

The remainder of this chapter is devoted to showing that there are very few fixed points of discontinuity of a martingale ξ such that $\xi(b)$ is L^1.

Let $\varepsilon > 0$. We say that a function $\xi \colon T \to M$, where $\langle M, \rho \rangle$ is a metric space (such as $\mathbf{L}^1$ or $\mathbf{R}$), is (nearly) ε-*continuous* at t in case whenever $t_1 \simeq t$ and $t_2 \simeq t$ we have $\rho(\xi(t_1), \xi(t_2)) \leq \varepsilon$; otherwise we say that ξ is (strongly) ε-*discontinuous* at t, and that t is a (strong) ε-*discontinuity* of ξ. Notice that if $s \simeq t$, then s is an ε-discontinuity of ξ if and only if t is. If t is an ε-discontinuity of ξ, then there are t_1 and t_2 infinitely close to t with $\rho(\xi(t_1), (t_2)) > \varepsilon$, so by the triangle inequality there is a t' (either t_1 or t_2) that is infinitely close to t with $\rho(\xi(t'), \xi(t)) > \varepsilon/2$. Notice also that t is a discontinuity of ξ if and only if t is an ε-discontinuity for some $\varepsilon \gg 0$.

Theorem 11.5 *Let ξ be a martingale such that $\xi(b)$ is L^1, and let $\varepsilon \gg 0$. Then there is a limited number of points $t_1, \ldots, t_n$, no two of which are infinitely close to each other, such that t is an ε-discontinuity of ξ in $\mathbf{L}^1$ if and only if $t \simeq t_i$ for some $i = 1, \ldots, n$.*

Proof. The set of all c such that $\|\xi(b) - \xi(b)^{(c)}\|_1 \leq \varepsilon/4$ contains all $c \simeq \infty$, and therefore contains some $c \ll \infty$. Let $\xi_c(t) = \mathbf{E}_t \xi(b)^{(c)}$. As remarked in Chapter 9, ξ_c is a martingale. Since conditional expectations reduce $\mathbf{L}^1$ norms, $\|\xi_c(t) - \xi(t)\|_1 \leq \varepsilon/4$ for all t.

Suppose that there are n points $t_1, \ldots, t_n$, no two of which are infinitely close to each other, with $\|\xi(s_i) - \xi(r_i)\|_1 > \varepsilon$, for some s_i and r_i with $s_i \simeq r_i \simeq t_i$. Choose notation so that $r_i < s_i$. Then the intervals $[r_i, s_i]$ are disjoint. By the triangle inequality, $\|\xi_c(s_i) - \xi_c(r_i)\|_1 > \varepsilon/2$, and so $\|\xi_c(s_i) - \xi_c(r_i)\|_2 > \varepsilon/2$. Since a martingale has orthogonal increments, we have

$$n\frac{\varepsilon^2}{4} \leq \sum_{i=1}^{n} \|\xi_c(s_i) - \xi_c(r_i)\|_2^2 \leq \|\xi_c(b) - \xi_c(a)\|_2^2$$

$$= \|\xi_c(b) - \mathbf{E}_a \xi_c(b)\|_2^2 \leq \|\xi_c(b)\|_2^2 \leq c^2.$$

Thus $n \leq 4c^2/\varepsilon^2 \ll \infty$. The proof is concluded by an appeal to the external least number principle. $\square$

*Theorem 11.6 *Let ξ be a martingale such that $\xi(b)$ is L^1. Then either:*

(i) *there is a limited number of points $t_1, \ldots, t_n$, no two of which are infinitely close to each other, such that t is a fixed point of discontinuity of ξ if and only if $t \simeq t_i$ for some i; or*

(ii) *there is an infinite sequence of points t_i, no two of which with standard indices are infinitely close to each other, such that t is a fixed point of discontinuity of ξ if and only if $t \simeq t_i$ for some standard i.*

Proof. By Theorem 11.4 (ii), t is a fixed point of discontinuity of ξ if and only if t is a $(1/k)$-discontinuity in $\mathbf{L}^1$ of ξ for some standard k. Let k be standard. By Theorem 11.5, there is a set

$$E_k = \{t_{k1}, \ldots, t_{kn_k}\},$$

with $n_k \ll \infty$ and $|t_{k_i} - t_{k-j}| \gg 0$ for $i \neq j$, such that the point t is a $(1/k)$-discontinuity in $\mathbf{L}^1$, but not a $(1/l)$-discontinuity in $\mathbf{L}^1$ for any natural number $l < k$, if and only if t is infinitely close to some (unique) element of E_k. By the sequence principle, there is a sequence $k \mapsto E_k$ of subsets of T such that these properties hold for all standard k. If there is a standard j such that E_k is empty for all standard $k > j$, let $t_1, \ldots, t_n$ be $\bigcup_{k \leq j} E_k$. Then $t_1, \ldots, t_n$ has all of the properties stated in (i). Otherwise, let j be nonstandard and let $t_1, \ldots, t_\nu$ be $\bigcup_{k \leq j} E_k$, with t_i defined arbitrarily for $i > \nu$. Then the sequence t_i has all of the properties stated in (ii). $\square$

Let us verify that (ii) really says that there are few fixed points of discontinuity.

Theorem 11.7 *Let T be a near interval. There does not exist a sequence t_i such that every element of T is infinitely close to some t_i with a standard index i.*

Proof. We argue by contradiction. Without loss of generality, assume that $a = 0$ and $b = 1$. Let x be in $[0,1]$ and let t be the largest element of the finite set T' such that $t \leq x$. Then $t \leq x \leq t + dt$, so that x is infinitely close to the element t of T and hence, by hypothesis, to some t_i with i standard. But Cantor's diagonal argument produces an x in $[0,1]$ that differs in the i'th decimal place from t_i; that is, $|x - t_i| \geq 10^{-i}$ for all i, including the standard ones. This is a contradiction. $\square$

We have seen that there are few fixed points of discontinuity in the sense of "cardinality". Now let us show that there are few fixed points of discontinuity in the sense of "measure".

*Theorem 11.8 *Let ξ be a martingale such that $\xi(b)$ is L^1, and let the function $f: \mathbf{R}^+ \to \mathbf{R}^+$ be such that $f(h) \simeq 0$ for all $h \simeq 0$. For all $\varepsilon \gg 0$ there are a natural number ν and intervals $[r_i, s_i]$, for $i = 1, \ldots, \nu$, such* *that we have $\sum_{i=1}^{\nu} f(s_i - r_i) \leq \varepsilon$ and each fixed point of discontinuity of ξ is in one of the intervals $[r_i, s_i]$.*

Proof. Let $\alpha > 0$ be infinitesimal. For all i in $\mathbf{N}$ let h_i be the largest integral multiple of α that is less than 1 and such that $f(h_i) \leq \varepsilon/2^i$, or $h_i = 0$ if no such number exists. Then for all $i \ll \infty$ we have $h_i \gg 0$ and $f(h_i) \leq \varepsilon/2^i$. Let t_i be the finite or infinite sequence given by Theorem 11.6, in case (i) let $\nu = n$, in case (ii) choose $\nu \simeq \infty$, and let $r_i = t_i - h_i/2$ and $s_i = t_i + h_i/2$. Then $\sum_{i=1}^{\nu} f(s_i - r_i) \leq \varepsilon$, and each fixed point of discontinuity of ξ is in one of the intervals $[r_i, s_i]$. $\square$

In particular, let ξ be a martingale indexed by a near interval T, such that $\xi(b)$ is L^1, and choose $f(h) = h$. Then we see that almost everywhere, with respect to pr_T ("normalized Lebesgue measure"), t is not a fixed point of discontinuity of ξ.

Finally, let us remark that if ξ is a martingale such that $\xi(b)$ is L^1, then $\xi(a) = \mathbf{E}_a \xi(b)$ is L^1, and so $\xi(b) - \xi(a)$ is L^1. Theorems 11.4–8 remain true under the hypothesis that $\xi(b) - \xi(a)$ is L^1, since ξ differs trivially from the martingale $\xi - \xi(a)$. In the next chapter we will study martingales under the weaker hypothesis that $\|\xi(b) - \xi(a)\|_1 \ll \infty$.

Chapter 12

Fluctuations of martingales

If ξ is either the Wiener walk or the Poisson walk, then $\mathbf{E}d\xi(r)^2 = dr$, so that $||\xi(t) - \xi(s)||_2^2 = |t - s|$. Thus in both examples the process is continuous in $\mathbf{L}^2$, and therefore in $\mathbf{L}^1$, for all t. By Theorem 11.3 (ii), for all t the process is continuous at t a.s. There are no fixed points of discontinuity. This does not imply, however, that a.s. the process is continuous at t for all t, because the exceptional sets may depend on t. Consider the Poisson walk indexed by a near interval. It takes only integer values, so the only way it could be continuous at all t would be to remain identically 0. But the probability of a jump at t is dt, so the probability of no jump is, by the independence of the increments, $\prod_{t\in T'}(1 - dt)$. But

$$\log \prod(1 - dt) = \sum \log(1 - dt) \sim -\sum dt = -(b - a)$$

by Theorem 5.3, since $\log(1 - dt) \sim -dt$. Therefore the probability of no jump is $\sim e^{-(b-a)}$, which is strongly less than 1. In the next chapter we shall see that the Wiener process is indeed a.s. continuous for all t. Typical trajectories of the two processes are sketched in Fig. 12.1.

This raises the question of how bad the discontinuities of a martingale trajectory can be. In conventional terms, for a bounded function there are two types of discontinuity: a discontinuity of the first kind (or jump discontinuity) and a discontinuity of the second kind (or oscillatory discontinuity), as in Fig. 12.2. (There are supposed to be infinitely many oscillations in Fig. 12.2 (b), as in $\sin \frac{1}{x}$ at the origin.) The notion of a function of limited fluctuation, defined in Chapter 6, is an external analogue of the internal notion of a function (with bounded domain) having jump discontinuities only. We may think of Fig. 12.2 (a) as illustrating a function of limited fluctuation, and (b) one of unlimited fluctuation.

The main result of this chapter is that under a very mild hypothesis, the trajectories of a supermartingale or submartingale are almost surely of

48

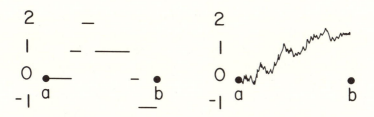

Poisson walk   Wiener walk

Figure 12.1: Two trajectories

limited fluctuation. First we establish a technical lemma, and then construct an investment strategy that leads to an estimate on the number of upcrossings of an interval.

Theorem 12.1 *Let ξ be a stochastic process indexed by a finite subset T of $\mathbf{R}$. Then ξ is of limited fluctuation a.s. if and only if for all $\varepsilon \gg 0$ and $k \simeq \infty$ we have*

$$\Pr\{\xi \text{ admits } k \text{ } \varepsilon\text{-fluctuations}\} \simeq 0.$$

Proof. The condition is clearly necessary. Suppose conversely that the condition holds. Let $M(\varepsilon, k)$ be the event that ξ admits k ε-fluctuations, and let $\delta \gg 0$. Let k_n be the least natural number such that

$$\Pr M \left(\frac{1}{n}, k_n\right) \leq \frac{\delta}{2^n},$$

a. First kind   b. Second kind

Figure 12.2: Discontinuities

and let

$$N = \bigcup_{n=1}^{\infty} M\left(\frac{1}{n}, k_n\right).$$

Then $\Pr N \leq \delta$. By hypothesis, k_n is limited when n is limited, so ξ is of limited fluctuation on N^c. Since $\delta \gg 0$ is arbitrary, this concludes the proof. $\square$

A speculator on the stock market hopes to make a killing from a wildly fluctuating market by following the "buy low–sell high" maxim. She decides to wait until the first time s_1 that the price of a share drops below λ_1 before buying a share, and then to sell it at the first subsequent time t_1 that the price rises above λ_2 (where $\lambda_2 > \lambda_1$). She then repeats the procedure as often as possible. The speculator's earnings at time t are given by

$$\varsigma_3(t) = \sum_{s<t} \eta_3(s) d\xi(s),$$

where η_3 is 1 when she is holding on to the stock (that is, when for some k it is true that $s_k \leq s$ but not $s \geq t_k$) and 0 at other times. Thus η_3 is a positive P-process. The speculator temporarily earns at least $\beta(\lambda_2 - \lambda_1)$ dollars, where β is the number of upcrossings of $[\lambda_1, \lambda_2]$ by ξ (that is, β is the largest k such that t_k is defined), but she may lose at the end if the last s_k is not followed by a t_k. However, this loss is at most $(\lambda_1 - \xi(b))^+$, where x^+ denotes $\max\{x, 0\}$. Thus

$$\varsigma_3(b) \geq \beta(\lambda_2 - \lambda_1) - (\lambda_1 - \xi(b))^+. \tag{12.1}$$

Suppose that ξ is a supermartingale, so that $0 \geq \mathbf{E}\varsigma_3(b)$. Then by (12.1) we have

$$(\lambda_2 - \lambda_1)\mathbf{E}\beta \leq \mathbf{E}(\lambda_1 - \xi(b))^+ \leq \|\xi(b)\|_1 + |\lambda_1|.$$

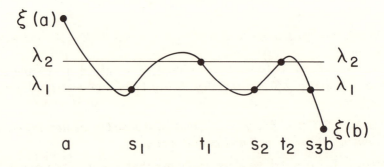

Figure 12.3: Two upcrossings

We have proved the following theorem.

Theorem 12.2 *Let ξ be a supermartingale, let $\lambda_1 < \lambda_2$, and let β be the number of upcrossings of $[\lambda_1, \lambda_2]$ by ξ. Then*

$$\mathbf{E}\beta \leq \frac{\|\xi(b)\|_1 + |\lambda_1|}{\lambda_2 - \lambda_1}.$$

Although we shall make no use of this fact, it is interesting that Theorem 12.2 also holds for a submartingale ξ. To see this, consider a conservative investor, who wants a sound stock for investment, and follows the opposite strategy of the speculator. That is, he buys a share at time a unless $\xi(a) \leq \lambda_1$. At time s_1 he sells and accepts his loss for tax purposes, waiting until time t_1 to buy again, and then repeats the procedure as often as necessary. His earnings at time t are given by

$$\varsigma_4(t) = \sum_{s<t} \eta_4(s) d\xi(s),$$

where $\eta_4(s) = 1 - \eta_3(s)$. Then η_4 is also a positive P-process. I claim that

$$\varsigma_4(b) \leq (\xi(b) - \lambda_1)^+ - \beta(\lambda_2 - \lambda_1). \tag{12.2}$$

To prove this, observe that if $\xi(a) \geq \lambda_1$, then he buys at the price $\xi(a)$ and loses at least $\lambda_2 - \lambda_1$ on each upcrossing, so that

$$\varsigma_4(b) \leq \xi(b) - \xi(a) - \beta(\lambda_2 - \lambda_1) \leq (\xi(b) - \lambda_1)^+ - \beta(\lambda_2 - \lambda_1).$$

If $\xi(a) < \lambda_1$ and $\beta = 0$, then he never buys, so that $\varsigma_4(b) = 0$ and (12.2) is trivially true. Finally, if $\xi(a) < \lambda_1$ and $\beta > 0$, then he first buys at a price $\geq \lambda_2$ after the first upcrossing, so that

$$\varsigma_4(b) \leq (\xi(b) - \lambda_2) - (\beta - 1)(\lambda_2 - \lambda_1) \leq (\xi(b) - \lambda_1)^+ - \beta(\lambda_2 - \lambda_1).$$

Thus (12.2) holds in all cases, and Theorem 12.2 also holds for a submartingale ξ (since then $0 \leq \mathbf{E}\varsigma_4(b)$).

Theorem 12.3 *Let ξ be a supermartingale or a submartingale with*

$$\|\xi(b) - \xi(a)\|_1 \ll \infty.$$

Then a.s. ξ is of limited fluctuation.

Proof. Since $-\xi$ is a supermartingale if and only if ξ is a submartingale, it is enough to prove the theorem for supermartingales. Also, there is no loss of generality in assuming that $\xi(a) = 0$. Let $\delta \gg 0$ and let

$$\lambda = \frac{2\|\xi(b)\|_1}{\delta},$$

so that $\lambda \ll \infty$. By Theorem 11.1,

$$\Pr\{\max|\xi(t)| \geq \lambda\} \leq \frac{2\|\xi(b)\|_1}{\lambda} \leq \delta. \qquad (12.3)$$

Now let $\varepsilon \gg 0$ be such that $n = \lambda/\varepsilon$ is an integer. Notice that $n \ll \infty$. Partition $[-\lambda, \lambda]$ into $2n$ subintervals of length ε.

Now suppose that we have $\max|\xi(t)| < \lambda$, and that the process ξ admits $2n + 2k$ 2ε-fluctuations, where k is an unlimited multiple of $2n$. Each 2ε-fluctuation produces either a downcrossing or an upcrossing of one of the $2n$ subintervals, so some subinterval has at least $1 + k/n$ crossings. Let β be the number of its upcrossings. The number of upcrossings and downcrossings of it can differ by at most 1, so $\beta \geq k/2n$. But

$$\Pr\left\{\beta \geq \frac{k}{2n}\right\} \leq \frac{2n}{k}\mathbf{E}\beta \leq \frac{n}{k\varepsilon}\left(\|\xi(b)\|_1 + \lambda\right)$$

by the Chebyshev inequality and Theorem 12.2. Since there are $2n$ subintervals, the probability that $\max|\xi(t)| < \lambda$ and that ξ admits $2n + 2k$ 2ε-fluctuations is

$$\leq \frac{2n^2}{k\varepsilon}\left(\|\xi(b)\|_1 + \lambda\right) \simeq 0.$$

Together with (12.3), this shows that

$$\Pr\{\xi \text{ admits } 2n + 2k \text{ } 2\varepsilon\text{-fluctuations}\} \lesssim \delta.$$

Since $\delta \gg 0$ is arbitrary, the proof is complete by Theorem 12.1. $\square$

Chapter 13

Discontinuities of martingales

We have just seen that, subject to a mild hypothesis, a.e. trajectory of a supermartingale or submartingale is of limited fluctuation. This gives some information on the nature of the possible discontinuities of a trajectory, but a finer analysis is called for. Let $\varepsilon \gg 0$. Then a.s. there is only a limited number of points with $|d\xi(t)| \geq \varepsilon$. Can it happen that two such points occur infinitely close together? Can it happen that each increment $d\xi(t)$ is infinitesimal but that there is a discontinuity (by means of an unlimited number of infinitesimal increments adding up to more than ε during an infinitesimal interval)? Can it happen that at some point t the trajectory is neither continuous from the left nor continuous from the right? Can it happen, in short, that there are two ε-fluctuations during some infinitesimal time interval? Certainly these local horrors can occur for a function of limited fluctuation, but we will see that the answer to these questions is *no* for a.e. trajectory of a supermartingale or submartingale ξ for which $0 \ll b - a$ and σ_ξ^2 is L^1 on $T' \times \Omega$. Recall that σ_ξ^2 is L^1 on $T' \times \Omega$ if each $\sigma_\xi^2(t)$ is L^1 on Ω (see the remark after the proof of Theorem 8.4).

If ξ is a stochastic process indexed by a finite subset T of $\mathbf{R}$, define the *proper time* τ_ξ by

$$\tau_\xi(t) = \sum_{s<t} \sigma_\xi^2(s)ds.$$

The proper time increases on each trajectory, but the rate at which it increases may vary from trajectory to trajectory. If $s \leq t$, we call $\tau_\xi(t) - \tau_\xi(s)$ the *proper time duration* of the interval $[s, t]$.

The price of a share of a certain stock is a martingale ξ satisfying the inequality $\|\xi(b) - \xi(a)\|_1 \ll \infty$. An investor is anxious not to have his capital tied up for a long time (proper time, actually—the anticipation of a rapidly fluctuating price makes the time seem long to him), and he figures

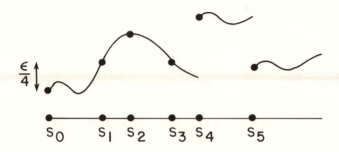

Figure 13.1: Another investment strategy

that a good time to buy is when the stock shows signs of activity. He chooses a natural number j, an $\varepsilon \gg 0$, and an infinitesimal $\alpha > 0$. Let $s_0 = a$, let s_1 be the first time such that $|\xi(s_1) - \xi(a)| \geq \varepsilon/4$, let s_2 be the first time with $s_2 > s_1$ such that $|\xi(s_2) - \xi(s_1)| \geq \varepsilon/4$, and so on.

The investor buys a share of the stock at time s_j and sells it at time s_{j+1} if $\tau_\xi(s_{j+1}) - \tau_\xi(s_j) \leq \alpha$; otherwise he sells it at the first time s such that $\tau_\xi(s + ds) - \tau_\xi(s_j) > \alpha$. He never buys again. His earnings at time t are given by

$$\varsigma_j(t) = \sum_{s<t} \eta_j(s) d\xi(s),$$

where η_j is a $\mathcal{P}$-process taking the values 0 and 1. (Notice that $\tau_\xi(s + ds) = \tau_\xi(s) + \sigma_\xi^2(s)ds$ is in $\mathcal{P}_s$, so that η_j is indeed a $\mathcal{P}$-process.)

We have

$$\|\varsigma_j(b)\|_2^2 = \mathbf{E}\left(\sum_{t\in T'} \eta_j(t) d\xi(t)\right)^2 = \mathbf{E}\sum_{t\in T'} \eta_j(t)^2 \sigma_\xi^2(t) dt \leq \alpha \simeq 0,$$

and by Chebyshev's inequality we have $\varsigma_j(b) \simeq 0$ a.s.

By Theorem 12.3, ξ is of limited fluctuation a.s., so that a.s. there is only a limited number of points s_j. Therefore

$$\max_j |\varsigma_j(b)| \simeq 0 \text{ a.s.} \tag{13.1}$$

Let $M(\varepsilon, \alpha)$ be the event that there are at least two ε-fluctuations within some interval of proper time duration $\leq \alpha$. Notice that if $|\xi(t_2) - \xi(t_1)| \geq \varepsilon$ and $|\xi(t_4) - \xi(t_3)| \geq \varepsilon$, where $t_1 \leq t_2 \leq t_3 \leq t_4$, then there are at least two s_j between t_1 and t_4, since there must be at least one in $[t_1, t_2]$ and at least one in $[t_3, t_4]$. Therefore, if $M(\varepsilon, \alpha)$ occurs, then for some j we have $\tau_\xi(s_{j+1} - \tau_\xi(s_j)) \leq \alpha$, and consequently

$$\max_j |\varsigma_j(b)| \geq \frac{\varepsilon}{4}.$$

Therefore, by (13.1), if $\varepsilon \gg 0$ and $\alpha \simeq 0$, then $\Pr M(\varepsilon, \alpha) \simeq 0$.

Let $\delta \gg 0$, and let α_n be the largest number (in the finite set of all possible values of $\tau_\xi(t) - \tau_\xi(s)$) such that

$$\Pr M\left(\frac{1}{n}, \alpha_n\right) \leq \frac{\delta}{2^n}.$$

Then $\alpha_n \gg 0$ for $n \ll \infty$, and if

$$N = \bigcup_{n=1}^{\infty} M\left(\frac{1}{n}, \alpha_n\right),$$

then $\Pr N \leq \delta$. But on N^c there do not exist two ε-fluctuations in any interval of infinitesimal proper time duration, for any $\varepsilon \gg 0$. Since $\delta \gg 0$ is arbitrary, we have proved the following result.

Theorem 13.1 *Let ξ be a martingale with $||\xi(b) - \xi(a)||_1 \ll \infty$. Then a.s. for all $\varepsilon \gg 0$ there do not exist two ε-fluctuations in any interval of infinitesimal proper time duration.*

Notice that for a normalized martingale ξ, the proper time differs trivially from the ordinary time: $\tau_\xi(t) = t - a$; and that

$$||\xi(b) - \xi(a)||_1 \leq ||\xi(b) - \xi(a)||_2 = \sqrt{b - a},$$

so that $||\xi(b) - \xi(a)||_1 \ll \infty$ if $b - a \ll \infty$. Then the conclusion of Theorem 13.1 holds. If each increment $d\xi$ is infinitesimal, everywhere on $T' \times \Omega$, it follows that a.e. trajectory is continuous at all times, for if $|\xi(t_2) - \xi(t_1)| \gg 0$ then there must be a t between t_1 and t_2 with $|\xi(t_2) - \xi(t)| \gg 0$ and $|\xi(t) - \xi(t_1)| \gg 0$. Thus Theorem 13.1 has the following corollary.

Corollary. *Almost every trajectory of the Wiener walk is continuous at all t.*

Let us convert the information about proper time in Theorem 13.1 into information about ordinary time.

Theorem 13.2 *Let ξ be a stochastic process with $b - a \gg 0$ such that σ_ξ^2 is L^1 on $T' \times \Omega$. Then a.s. the proper time τ_ξ is absolutely continuous.*

In particular, a.s. every interval of infinitesimal length is of infinitesimal proper time duration.

Proof. By the Fubini theorem, for a.e. ω the function $t \mapsto \sigma_\xi^2(t)$ is L^1 on T'. By the Lebesgue theorem, this implies that a.s. τ_ξ is absolutely continuous (since $d\tau_\xi(t)/dt = \sigma_\xi^2(t)$ and $b - a \gg 0$). The final statement of the theorem is merely that an absolutely continuous function is continuous. $\square$

Let T be a finite subset of $\mathbf{R}$ and let $\xi: T \to \mathbf{R}$. We say that s in T' is an *ε-jump* of ξ in case $|d\xi(s)| \geq \varepsilon$, and a *jump* of ξ in case it is an ε-jump for some $\varepsilon \gg 0$. If s is a jump of ξ with $ds \simeq 0$, and if $t \simeq s$, then t is a discontinuity of ξ. We say that a point t is a *jump discontinuity* of ξ in case there is a unique jump s with $s \simeq t$.

Theorem 13.3 *Let T be a finite subset of $\mathbf{R}$ and let $\xi: T \to \mathbf{R}$ be such that for all $\varepsilon \gg 0$ there do not exist two ε-fluctuations in any infinitesimal interval. Then every discontinuity t of ξ is a jump discontinuity. If s is the jump with $s \simeq t$, then ξ is continuous from the left at t if and only if $t \leq s$, and ξ is continuous from the right at t if and only if $t > s$. If all of the increments of ξ are infinitesimal, then ξ is continuous. No two jumps of ξ are infinitely close to each other. If $b - a \ll \infty$, then for all $\varepsilon \gg 0$ there is only a limited number of ε-jumps.*

Proof. Let t be a discontinuity of ξ. That is, there are t_1 and t_2 infinitely close to t, with $t_1 \leq t_2$, such that for some $\delta \gg 0$ we have $|d\xi(t_2) - d\xi(t_1)| \geq \delta$. Let s be the first time $\geq t_1$ such that $|\xi(s + ds) - \xi(t_1)| \geq \delta/2$. By the hypothesis on ξ, we have $|\xi(s+ds) - \xi(t_1)| \gtrsim \delta$ and $|\xi(s) - \xi(t_1)| \simeq 0$. Therefore $|\xi(s + ds) - \xi(s)| \gtrsim \delta$, so that s is a $(\delta/2)$-jump. Thus there is a jump s with $s \simeq t$, it is clearly unique, and $ds \leq t_2 - t_1 \simeq 0$, so that t is a jump discontinuity. The remaining statements in the theorem are now obvious. $\square$

By Theorems 13.1–3 we have the following result.

Theorem 13.4 *Let ξ be a martingale with $0 \ll b - a$ such that σ_ξ^2 is L^1 on $T' \times \Omega$. Then the conclusions of Theorem 13.3 hold for a.e. trajectory.*

Using the sequence principle, we can also show (under these hypotheses and the assumption that $b - a \ll \infty$) that a.s. there is a sequence of points s_i in T such that for all s in T, s is a jump of ξ if and only if $s = s_i$ for some standard i.

Chapter 14

The Lindeberg condition

We say that a stochastic process ξ satisfies the (near) *Lindeberg condition* in case for all $\varepsilon \gg 0$ we have

$$\mathbf{E} \sum d\xi(t)^2 \simeq \mathbf{E} \sum d\xi(t)^{(\varepsilon)2}. \tag{14.1}$$

For a normalized martingale, the left hand side of (14.1) is equal to $b - a$. The Wiener walk satisfies the Lindeberg condition because the two sides of (14.1) are equal, but for the Poisson walk the right hand side is 0 for $\varepsilon < 1$.

One way to obtain a normalized martingale is this. Let $x_1, \ldots, x_\nu$, with $\nu \simeq \infty$, be independent random variables of mean 0 with $0 < \mathbf{E}x_k^2 \ll \infty$ for $1 \le k \le \nu$, and let

$$s_n = \sqrt{\sum_{k=1}^{n} \mathbf{E}x_k^2}.$$

Suppose that $s_\nu \simeq \infty$. Let T be the set of all numbers of the form s_n^2/s_ν^2 for $n = 0, \ldots, \nu$. Then T is a near interval with $a = 0$ and $b = 1$. For t in T, with $t = s_n^2/s_\nu^2$, let

$$\xi(t) = \frac{1}{s_\nu} \sum_{k=1}^{n} x_k.$$

Then ξ is a normalized martingale, which we will call the normalized martingale *associated* with $x_1, \ldots, x_\nu$. It satisfies the Lindeberg condition if and only if for all $\varepsilon \gg 0$,

$$\frac{1}{s_\nu^2} \sum_{k=1}^{\nu} \mathbf{E}x_k^{(\varepsilon s_\nu)2} \simeq 1. \tag{14.2}$$

The relation (14.2) is very similar to the conventional form of the Lindeberg condition, but it is much easier to understand when expressed in the form (14.1) for the associated normalized martingale.

Theorem 14.1 *Let ξ be a stochastic process satisfying the Lindeberg condition. Then a.s. each increment is infinitesimal.*

Proof. For $\varepsilon \gg 0$ we have

$$0 \simeq \mathbf{E} \sum \left(d\xi(t)^2 - d\xi(t)^{(\varepsilon)2} \right) = \sum_t \sum_{|\lambda| > \varepsilon} \lambda^2 \mathrm{pr}_{d\xi(t)}(\lambda)$$

$$\geq \varepsilon^2 \sum_t \mathrm{Pr}\{|d\xi(t)| > \varepsilon\} \geq \varepsilon^2 \mathrm{Pr}\{\max|d\xi(t)| > \varepsilon\}.$$

Since $\varepsilon \gg 0$, it follows that $\mathrm{Pr}\{\max|d\xi(t)| > \varepsilon\} \simeq 0$, and since $\varepsilon \gg 0$ is arbitrary, this means by Theorem 7.1 that $\max|d\xi(t)| \simeq 0$ a.s. $\square$

Theorem 14.2 *Let ξ be a martingale with $0 \ll b - a$ such that σ_ξ^2 is L^1 on $T' \times \Omega$, that satisfies the Lindeberg condition. Then a.s. ξ is continuous on T.*

Proof. By Theorems 14.1 and 13.4. $\square$

If $\xi : T \to \mathbf{R}$, we define

$$q_\xi(t) = \sum_{s < t} d\xi(s)^2.$$

We call q_ξ the *quadratic variation* of ξ. If T is a near interval and ξ is a smooth function, then the quadratic variation will be infinitesimal. But for a martingale ξ we have $\mathbf{E}q_\xi(t) = \|\xi(t) - \xi(a)\|_2^2$, so that it is quite normal to have non-infinitesimal quadratic variation.

Recall the typical trajectories of the Poisson and Wiener walks illustrated in Fig. 12.1. The corresponding quadratic variations are graphed in Fig. 14.1. Notice that for the Wiener walk, but not for the Poisson walk, the quadratic variation does not depend on the trajectory. To show that this phenomenon results from the Lindeberg condition, we first establish a truncation lemma similar to Theorem 10.3.

Theorem 14.3 *Let ξ be a martingale that satisfies the Lindeberg condition. For $\alpha > 0$, let ξ_α be the martingale with $\xi_\alpha(a) = \xi(a)$ and*

$$d\xi_\alpha(t) = d\xi(t)^{(\alpha)} - \mathbf{E}_t d\xi(t)^{(\alpha)},$$

so that $|d\xi_\alpha(t)| \leq 2\alpha$ for all t and ω. Then there is an infinitesimal α such that a.s. for all t we have $\xi_\alpha(t) \simeq \xi(t)$, $q_{\xi_\alpha}(t) \simeq q_\xi(t)$, and $\tau_{\xi_\alpha} \simeq \tau_\xi(t)$.

Proof. We have $\max|d\xi(t)| \simeq 0$ a.s. by Theorem 14.1, so it follows from Theorem 7.1 that for any sufficiently large infinitesimal α we have $\mathrm{Pr}\{\max|d\xi(t)| \geq \alpha\} \simeq 0$. Then a.s. we have

$$\xi(t) = \xi(a) + \sum_{s < t} d\xi(s)^{(\alpha)}.$$

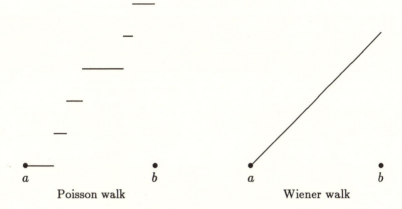

Poisson walk   Wiener walk

Figure 14.1: Quadratic variation

Therefore, to show that a.s. $\xi_\alpha(t) \simeq \xi(t)$ for all t, it is enough to show that

$$\sum \left| \mathbf{E}_t d\xi(t)^{(\alpha)} \right| \simeq 0 \text{ a.s.,}$$

and for this it suffices to show that its expectation is infinitesimal. But since $\mathbf{E}_t d\xi(t) = 0$ we have

$$\mathbf{E} \sum \left| \mathbf{E}_t d\xi(t)^{(\alpha)} \right| = \mathbf{E} \sum \left| \mathbf{E}_t \left(d\xi(t) - d\xi(t)^{(\alpha)} \right) \right|$$

$$\leq \mathbf{E} \sum \mathbf{E}_t \left| d\xi(t) - d\xi(t)^{(\alpha)} \right| \leq \frac{1}{\alpha} \mathbf{E} \sum \left(d\xi(t)^2 - d\xi(t)^{(\alpha)2} \right).$$

If $\alpha \gg 0$, this in infinitesimal by the Lindeberg condition. Hence the set of all α such that $\mathbf{E} \sum \left| \mathbf{E}_t d\xi(t)^{(\alpha)} \right| \leq \alpha$ contains all $\alpha \gg 0$ and thus, by overspill, contains all sufficiently large infinitesimal α.

By the same reasoning, $\sum \left(d\xi(t)^2 - d\xi(t)^{(\alpha)2} \right) \simeq 0$ a.s. But $d\xi_\alpha(t)^2$ differs from $d\xi(t)^{(\alpha)2}$ by

$$-2d\xi(t)^{(\alpha)} \mathbf{E}_t d\xi(t)^{(\alpha)} + \left(\mathbf{E}_t d\xi(t)^{(\alpha)} \right)^2,$$

which in absolute value is less than $3\alpha \left| \mathbf{E}_t d\xi(t)^{(\alpha)} \right|$. Consequently, a.s. $q_{\xi_\alpha}(t) \simeq q_\xi(t)$ for all t. Finally, apply the same reasoning with $\mathbf{E}_t$ preceding each term and conclude that a.s. $\tau_{\xi_\alpha}(t) \simeq \tau_\xi(t)$ for all t. $\square$

Theorem 14.4 *Let ξ be a martingale that satisfies the Lindeberg condition, with $\|\xi(b) - \xi(a)\|_2^2 \ll \infty$. Then a.s. $q_\xi(t) \simeq \tau_\xi(t)$ for all t.*

Proof. By the previous theorem, we may assume that for ξ itself there is an $\alpha \simeq 0$ such that $|d\xi(t)| \leq \alpha$ for all t and ω. Observe that $q_\xi - \tau_\xi$ is a martingale with initial value 0. We have

$$\|q_\xi(b) - \tau_\xi(b)\|_1^2 \leq \|q_\xi(b) - \tau_\xi(b)\|_2^2 = \sum \|dq_\xi(t) - d\tau_\xi(t)\|_2^2$$

$$= \sum \left\| d\xi(t)^2 - \mathbf{E}_t d\xi(t)^2 \right\|_2^2 \leq \sum \left\| d\xi(t)^2 \right\|_2^2 = \sum \mathbf{E} d\xi(t)^4$$

$$\leq \alpha^2 \sum \mathbf{E} d\xi(t)^2 = \alpha^2 \|\xi(b) - \xi(a)\|_2^2 \simeq 0.$$

By Theorem 11.1 we have $\Pr\{\max |q_\xi(t) - \tau_\xi(t)| \geq \lambda\} \simeq 0$ for all $\lambda \gg 0$, so by Theorem 7.1 we have $\max |q_\xi(t) - \tau_\xi(t)| \simeq 0$ a.s. $\square$

Chapter 15

The maximum of a martingale

Let us return to the proof of Theorem 11.1. If ξ is a submartingale, then, by (11.3),

$$0 \leq \mathbf{E}(\xi(b) - \xi(a) - \lambda)\chi_{\{\max(\xi(t) - \xi(a)) \geq \lambda\}}.$$

We can relativize this result to the algebra $\mathcal{P}_a$, replacing $\mathbf{E}$ by $\mathbf{E}_a$ and λ by any element of $\mathcal{P}_a$. In particular, if we replace λ by $\lambda - \xi(a)$, we obtain

$$0 \leq \mathbf{E}_a(\xi(b) - \lambda)\chi_{\{\max \xi(t) \geq \lambda\}}.$$

Taking the absolute expectation, we have

$$\lambda \Pr\{\max \xi(t) \geq \lambda\} \leq \mathbf{E}\xi(b)\chi_{\{\max \xi(t) \geq \lambda\}} \leq \|\xi(b)\|_1. \qquad (15.1)$$

We have proved the following theorem.

Theorem 15.1 *Let ξ be a submartingale. Then* (15.1) *holds.*

Here is another proof of this theorem. Let $\Lambda = \{\max \xi(t) \geq \lambda\}$, and let

$$\Lambda_t = \{\xi(t) \geq \lambda \text{ and } \xi(s) < \lambda \text{ for all } s < t\}.$$

Then Λ is the disjoint union of the Λ_t, and $\chi_{\Lambda_t} \in \mathcal{P}_t$. We have

$$\|\xi(b)\|_1 \geq \mathbf{E}\xi(b)\chi_\Lambda = \mathbf{E}\sum \xi(b)\chi_{\Lambda_t} = \mathbf{E}\sum \mathbf{E}_t \xi(b)\chi_{\Lambda_t}$$

$$= \mathbf{E}\sum \chi_{\Lambda_t} \mathbf{E}_t \xi(b) \geq \mathbf{E}\sum \chi_{\Lambda_t}\xi(t) \geq \mathbf{E}\sum \chi_{\Lambda_t}\lambda = \lambda \Pr \Lambda,$$

which proves (15.1).

If ξ is a martingale, then $|\xi|^p$, for $1 \leq p < \infty$, is a submartingale, by Theorem 11.2. Since

$$\{\max |\xi(t)| \geq \lambda\} = \{\max |\xi(t)|^p \geq \lambda^p\}$$

for $\lambda > 0$, Theorem 11.1 implies that

$$\Pr\{\max|\xi(t)| \geq \lambda\} \leq \frac{1}{\lambda^p}\mathbf{E}|\xi(b)|^p. \qquad (15.2)$$

If $\max|\xi(t)|$ had an $\mathbf{L}^p$ norm smaller than that of $\xi(b)$, then (15.2) would follow from the Chebyshev inequality. This is false in general, but we do have the remarkable result that

$$\|\max|\xi(t)|\|_p \leq p'\|\xi(b)\|_p \qquad (15.3)$$

where $1 < p \leq \infty$ and p' is the conjugate exponent. This is a consequence of Theorem 15.1 and the following result.

Theorem 15.2 *Let x and y be positive random variables such that for all $\lambda > 0$ we have*

$$\Pr\{y \geq \lambda\} \leq \frac{1}{\lambda}\mathbf{E}x\chi_{\{y \geq x\}}.$$

Then for all p with $1 < p \leq \infty$ we have $\|y\|_p \leq p'\|x\|_p$.

Proof. For $1 < p < \infty$ we have

$$\|y\|_p^p = \mathbf{E}y^p = \sum \lambda^p \mathrm{pr}_y(\lambda) = -\int_0^\infty \lambda^p d\Pr\{y \geq \lambda\}$$

$$= \int_0^\infty p\lambda^{p-1}\Pr\{y \geq \lambda\}d\lambda \leq \int_0^\infty p\lambda^{p-2}\mathbf{E}x\chi_{\{y \geq \lambda\}}d\lambda$$

$$= \mathbf{E}\left(x\int_0^y p\lambda^{p-2}d\lambda\right) = \frac{p}{p-1}\mathbf{E}xy^{p-1} \leq p'\|x\|_p\|y^{p-1}\|_{p'}$$

$$= p'\|x\|_p\|y^{p/p'}\|_{p'} = p'\|x\|_p\|y\|_p^{p-1}.$$

The case $p = \infty$ follows by letting $p \to \infty$. $\square$

Chapter 16

The law of large numbers

For some reason, probabilists use the the phrase "law of large numbers" instead of the more descriptive "law of averages". But the problem concerns the behavior of the averages $y_n = (x_1 + \cdots + x_n)/n$ of a sequence of random variables.

Let $x_1, \ldots, x_\nu$ be independent random variables of mean 0 and variance 1, let P_n be the algebra generated by the random variables $x_1, \ldots, x_n$, and let $y_n = (x_1 + \cdots + x_n)/n$. Then

$$dy_n = \frac{x_1 + \cdots + x_n + x_{n+1}}{n+1} - \frac{x_1 + \cdots + x_n}{n} = -\frac{y_n}{n+1} + \frac{x_{n+1}}{n+1}. \quad (16.1)$$

Since the x_n are independent, we have $\mathbf{E}_n x_{n+1} = \mathbf{E} x_{n+1} = 0$. Therefore

$$Dy_n = -\frac{y_n}{n+1} \quad \text{and} \quad d\hat{y}_n = \frac{x_{n+1}}{n+1}.$$

For $n \simeq \infty$ we have $\|y_n\|_2^2 = 1/n \simeq 0$, so that by the Chebyshev inequality $y_n \simeq 0$ a.s. That is, $y_1, \ldots, y_\nu$ converges to 0 in probability. This is the weak law of large numbers. We can also prove the strong law of large numbers: $y_1, \ldots, y_\nu$ converges to 0 a.s. To see this, notice that

$$\|Dy_n\|_1 \leq \|Dy_n\|_2 = \frac{1}{n+1} \frac{1}{\sqrt{n}},$$

so that $\sum \|Dy_n\|_1$ converges. By Theorem (10.1) (ii), the predictable process associated to y_n converges a.s. Also,

$$\sum \|d\hat{y}_n\|_2^2 = \sum \frac{1}{(n+1)^2}$$

converges, so that by Theorem 11.3 (i) the martingale associated to y_n converges a.s. Thus y_n converges a.s., and by the weak law of large numbers it converges to 0 a.s. This establishes the strong law of large numbers for the case discussed at the beginning of Chapter 4, but we shall prove some stronger results. Notice that the proof did not really use independence, but only the fact that the x_n are the increments of a martingale.

Theorem 16.1 *Let $x_1, \ldots, x_\nu$ be the increments of a martingale.*

(i) *If $\sum_{i=1}^{\nu} \dfrac{\mathbf{E}x_i^2}{i^2}$ converges, then $(x_1 + \cdots + x_n)/n$ converges to 0 a.s.*

(ii) *If $\sum_{i=1}^{\nu} \dfrac{\mathbf{E}x_i^2}{i^2} \ll \infty$, then $(x_1 + \cdots + x_n)/n$ is of limited fluctuation a.s.*

Proof. Let

$$y_n = \frac{1}{n}\sum_{i=1}^{n} x_i \ \text{ and } \ z_n = \sum_{i=1}^{n} \frac{x_i}{i}.$$

Then

$$y_n = -\frac{1}{n}\sum_{j=1}^{n} z_j + \frac{n+1}{n}z_n, \tag{16.2}$$

as can be seen by collecting coefficients of x_i on both sides. Now the z_n are a martingale with $z_0 = 0$ and

$$\|z_\nu\|_2^2 = \sum_{i=1}^{\nu} \frac{\mathbf{E}x_i^2}{i^2} \ll \infty,$$

so Theorem 11.1 implies that a.s. z_n is limited for all n. In case (i), Theorem 11.3 (i) implies that z_n converges a.s., and in case (ii), Theorem 12.3 implies that z_n is of limited fluctuation a.s. Therefore the proof will be concluded if we can establish the following two claims:

(I) *If z_n is limited and convergent, then y_n converges to 0.*

(II) *If z_n is limited and of limited fluctuation, then y_n is of limited fluctuation.*

To prove (I), let $n \simeq \infty$. Since z_n is limited,

$$\frac{n+1}{n}z_n \simeq z_n \simeq z_\nu.$$

Let $\varepsilon \gg 0$ and let k be the largest natural number $\leq \nu$ such that $|z_k - z_\nu| > \varepsilon$ (and $k = 0$ if there is no such number). Then $k \ll \infty$, each $|z_j| \ll \infty$, but $n \simeq \infty$, so that

$$\frac{1}{n}\sum_{j=1}^{k} z_j \simeq 0$$

and thus

$$\left| \frac{1}{n} \sum_{j=1}^{n} z_j - z_\nu \right| < \varepsilon.$$

Since $\varepsilon \gg 0$ is arbitrary, this means that the right hand side of (16.2) is infinitely close to $-z_\nu + z_\nu = 0$, which proves (I).

To prove (II), observe that $z_n/n \simeq 0$ for $n \simeq \infty$, so that z_n/n converges to 0 and is therefore of limited fluctuation. By the triangle inequality, the sum of two sequences of limited fluctuation is of limited fluctuation. Therefore

$$\frac{n+1}{n} z_n = z_n + \frac{1}{n} z_n$$

is of limited fluctuation. Let

$$w_n = \frac{1}{n} \sum_{j=1}^{n} z_j.$$

It remains to show that w_n is of limited fluctuation. But, as in the proof of Theorem 12.3, we see that if w_n is of unlimited fluctuation, then there exist $\lambda_1 \ll \lambda_2$ such that w_n has an unlimited number of upcrossings of $[\lambda_1, \lambda_2]$. Define $s_1, t_1, s_2, t_2, \ldots$ for the sequence w_n as in Chapter 12 (see Fig. 12.3). Now in order for w_n to upcross $[\lambda_1, \lambda_2]$ between s_1 and t_1, there must be a z_{j_1}, with j_1 between s_1 and t_1, such that $z_{j_1} \geq \lambda_2$. Similarly, for it to upcross again at a later time, there must be a z_{k_1}, with k_1 between t_1 and s_2, such that $z_{k_1} \leq \lambda_1$, and so forth. Thus z_n must have at least $\beta - 1$ upcrossings of $[\lambda_1, \lambda_2]$ if w_n has β upcrossings of $[\lambda_1, \lambda_2]$. But this is impossible for $\beta \simeq \infty$, since z_n is of limited fluctuation. This proves (II). $\square$

Here is an example in which case (ii) applies but not case (i). Let $x_n = 0$ for $n < \nu$, and let $x_\nu = \nu$ with probability $\frac{1}{2}$ and $x_\nu = -\nu$ with probability $\frac{1}{2}$, so that $\sum \mathbf{E} x_i^2 / i^2 = 1$. Then the averages y_n are 0 for $n < \nu$, but $y_\nu = \pm 1$, each with probability $\frac{1}{2}$.

Thus the y_n with n unlimited need not be infinitesimal in case (ii). Nevertheless, the y_n with n limited become and remain smaller than any $\varepsilon \gg 0$. To see this, observe that by Theorem 6.1, if

$$\sum_{i=1}^{\nu} \frac{\mathbf{E} x_i^2}{i^2} \ll \infty,$$

then there is an unlimited μ with $\mu \leq \nu$ such that

$$\sum_{i=1}^{\mu} \frac{\mathbf{E} x_i^2}{i^2} \text{ converges,}$$

and so case (i) applies to $x_1, \ldots, x_\mu$.

Theorem 16.1 generalizes to other exponents p, but the proof consists of a truncation argument that reduces the situation to the case $p = 2$:

Theorem 16.2 *Let* $x_1, \ldots, x_\nu$ *be the increments of a martingale, and let* $0 \ll p \leq 2$.

(i) *If* $\displaystyle\sum_{i=1}^{\nu} \frac{\mathbf{E}|x_i|^p}{i^p}$ *converges, then* $(x_1 + \cdots + x_n)/n$ *converges to* 0 *a.s.*

(ii) *If* $\displaystyle\sum_{i=1}^{\nu} \frac{\mathbf{E}|x_i|^p}{i^p} \ll \infty$, *then* $(x_1 + \cdots + x_n)/n$ *is of limited fluctuation a.s.*

Proof. In both cases, I claim that a.s. there is a $c \ll \infty$ such that $|x_n| \leq cn$ for all n. To see this, let $\varepsilon \gg 0$, let $a = \sum \mathbf{E}|x_i|^p/i^p$, and let $c = (a/\varepsilon)^{1/p}$. Then $c \ll \infty$, since $a \ll \infty$ and $p \gg 0$. By the Chebyshev inequality,

$$\Pr\{|x_n| > cn \text{ for some } n\} \leq \sum \Pr\{|x_n| > cn\}$$

$$\leq \sum \frac{\mathbf{E}|x_n|^p}{(cn)^p} = \frac{a}{c^p} = \varepsilon.$$

Since $\varepsilon \gg 0$ is arbitrary, this proves the claim.

Let us say that a sequence of random variables *behaves* in case it converges to 0 a.s. in case (i) and is of limited fluctuation a.s. in case (ii). By what we have just shown, it suffices to prove that, for $c \ll \infty$, the sequence $(x_1^{(c1)} + \cdots + x_n^{(cn)})/n$ behaves. Now the $x_n^{(cn)}$ are not in general the increments of a martingale, but the $x_n^{(cn)} - \mathbf{E}_{n-1}x_n^{(cn)}$ are. We have

$$\sum \left\| \mathbf{E}_{n-1}\frac{x_n^{(cn)}}{n} \right\|_1 = \sum \frac{1}{n} \left\| \mathbf{E}_{n-1}\left(x_n - x_n^{(cn)}\right) \right\|_1$$

$$= \sum \frac{1}{n} \sum_{|\lambda| > cn} |\lambda| \mathrm{pr}_{x_n}(\lambda) \leq \sum \sum_{|\lambda| > cn} \frac{|\lambda|^p}{c^{p-1}n^p} \mathrm{pr}_{x_n}(\lambda)$$

$$\leq \frac{1}{c^{p-1}} \sum \frac{\mathbf{E}|x_n|^p}{n^p}.$$

By Theorem 10.2,

$$\sum \frac{\mathbf{E}_{n-1}x_n^{(cn)}}{n}$$

converges a.s. in case (i) and is of limited fluctuation a.s. in case (ii), and in both cases a.s. all of the terms are limited. By (I) and (II) of the previous proof, this means that

$$\left(\mathbf{E}_0 x_1^{(c1)} + \cdots + \mathbf{E}_{n-1}x_n^{(cn)}\right)/n$$

behaves, so we need only show that the averages of

$$x_n^{(cn)} - \mathbf{E}_{n-1} x_n^{(cn)}$$

behave. To do this we need only verify the hypotheses of Theorem 16.1:

$$\sum \frac{\mathbf{E}\left(x_n^{(cn)} - \mathbf{E}_{n-1} x_n^{(cn)}\right)^2}{n^2} \leq \sum \frac{\mathbf{E} x_n^{(cn)2}}{n^2}$$

$$= \sum \frac{1}{n^2} \sum_{|\lambda| \leq cn} |\lambda|^2 \mathrm{pr}_{x_n}(\lambda),$$

and since $p \leq 2$, this is

$$\leq \sum \frac{1}{n^2} \sum_{|\lambda| \leq cn} \frac{|\lambda|^p}{(cn)^{p-2}} \mathrm{pr}_{x_n}(\lambda)$$

$$= \frac{1}{c^{p-2}} \sum \frac{1}{n^p} \sum_{|\lambda| \leq cn} |\lambda|^p \mathrm{pr}_{x_n}(\lambda)$$

$$\leq \frac{1}{c^{p-2}} \sum \frac{\mathbf{E}|x_n|^p}{n^p}. \quad \square$$

Corollary. *Let $x_1, \ldots, x_\nu$ be the increments of a martingale, and let $p \gg 1$. If each $\|x_n\|_p$ is limited, then $(x_1 + \cdots + x_n)/n$ converges to 0 a.s.*

Proof. There is no loss of generality in assuming that $p \leq 2$. Let $M = \max \|x_n\|_p$. Then $M \ll \infty$, since $M = \|x_n\|_p$ for some n. Therefore the hypothesis of Theorem 16.2 (i) is satisfied. $\square$

Notice that Theorem 16.2 is trivial for $p = 1$: the conclusions follow by Theorem 10.2 and (I) and (II), even without the assumption that the x_n are the increments of a martingale.

But the corollary is false for $p = 1$, as the following counterexample shows. Let the x_n, for $n = 3, 4, \ldots, \nu$, be independent and satisfy

$$x_n = \begin{cases} n & \text{with probability } \frac{1}{2} \frac{1}{n \log n}, \\ 0 & \text{with probability } \frac{1}{2} - \frac{1}{2} \frac{1}{n \log n}, \\ -\frac{1}{\log n} & \text{with probability } \frac{1}{2}. \end{cases}$$

Then $\mathbf{E} x_n = 0$ and $\|x_n\|_1 = 1/\log n$. Notice that each x_n is L^1 and $\|x_n\|_1$ is even infinitesimal for $n \simeq \infty$. But by the cardinal version of the Borel-Cantelli theorem (Theorem 7.4 (ii)), a.s. $x_n = n$ for an unlimited number of n. Therefore a.s.

$$y_n = \frac{x_3 + \cdots + x_n}{n - 2} \geq 1 - \frac{1}{\log 3}$$

for an unlimited number of n, and so does not converge to 0. In fact, using (16.1) one can easily show that a.s. y_n is of unlimited fluctuation.

The following theorem shows that there is an important case, that of repeated independent observations of a random variable, in which the strong law of large numbers does hold with L^1 hypotheses.

Theorem 16.3 *Let x be a random variable and let $x_1, \ldots, x_\nu$ be independent random variables with the same probability distribution as x.*

(i) *If x is L^1, then $(x_1 + \cdots + x_n)/n$ converges to $\mathbf{E}x$ a.s.*

(ii) *If $\|x\|_1 \ll \infty$, then $(x_1 + \cdots + x_n)/n$ is of limited fluctuation a.s.*

Proof. The strategy of the proof is the same as for Theorem 16.2, but some of the tactics are different. There is no loss of generality in assuming that $\mathbf{E}x = 0$.

I claim that a.s. there is a $c \ll \infty$ such that $|x_n| \leq cn$ for all n. To see this, let $\varepsilon \gg 0$, let $c = \mathbf{E}|x|/\varepsilon$, let

$$E_k = \{ck < |x| \leq c(k+1)\}$$

and observe that

$$\Pr\{|x_n| > cn \text{ for some } n\} \leq \sum_{n=1}^{\nu} \Pr\{|x_n| > cn\}$$

$$= \sum_{n=1}^{\nu} \Pr\{|x| > cn\} = \sum_{n=1}^{\nu} \sum_{k=n}^{\infty} \Pr E_k$$

$$= \sum_{k=0}^{\infty} \sum_{n=1}^{\min\{\nu,k\}} \Pr E_k \leq \sum_{k=0}^{\infty} k \Pr E_k$$

$$\leq \frac{1}{c} \sum_{\lambda} \mathrm{pr}_x(\lambda) = \frac{1}{c} \mathbf{E}|x| = \varepsilon,$$

which proves the claim.

Therefore it is enough to show that, for $c \ll \infty$, the averages of $x_n^{(cn)}$ behave. Now $\mathbf{E}x_n^{(cn)} = \mathbf{E}x^{(cn)} = \mathbf{E}\left(x - x^{(cn)}\right)$, and it is smaller in absolute value than

$$\sum_{|\lambda| > cn} |\lambda| \mathrm{pr}_x(\lambda).$$

Also, for $n < m$ we have

$$\left|\mathbf{E}x_n^{(cn)} - \mathbf{E}x_m^{(cm)}\right| = \left|\mathbf{E}x^{(cn)} - \mathbf{E}x^{(cm)}\right| \leq \sum_{cn < |\lambda| \leq cm} |\lambda| \mathrm{pr}_x(\lambda).$$

Thus $\mathbf{E}x_n^{(cn)}$ is always limited, and it behaves. If follows, as in the proof of (I) and (II), that the averages of $\mathbf{E}x_n^{(cn)}$ behave. (We have exploited independence in this argument—it was essential that we were dealing with absolute expectations rather than conditional expectations.) Therefore we need only show that $x_n^{(cn)} - \mathbf{E}x_n^{(cn)}$ satisfies the hypotheses of Theorem 16.1. But

$$\sum_{n=m}^{\nu} \frac{\mathbf{E}\left(x_n^{(cn)} - \mathbf{E}x_n^{(cn)}\right)^2}{n^2} \leq \sum_{n=m}^{\nu} \frac{\mathbf{E}x^{(cn)2}}{n^2}$$

$$= \sum_{n=m}^{\nu} \frac{1}{n^2} \sum_{|\lambda| \leq cn} \lambda^2 \mathrm{pr}_x(\lambda) = \sum_{|\lambda|} \lambda^2 \sum_{n \geq \max\{m, |\lambda|/c\}} \frac{1}{n^2}$$

$$\leq \sum_{|\lambda| \leq \sqrt{m}} \lambda^2 \mathrm{pr}_x(\lambda) \sum_{n=m}^{\nu} \frac{1}{n^2} + \sum_{|\lambda| > \sqrt{m}} \lambda^2 \mathrm{pr}_x(\lambda) \sum_{n=|\lambda|/c}^{\nu} \frac{1}{n^2}$$

$$\leq \sum_{|\lambda| \leq \sqrt{m}} \lambda^2 \frac{1}{m-1} + \sum_{|\lambda| > \sqrt{m}} \lambda^2 \mathrm{pr}_x(\lambda) \frac{1}{(|\lambda|/c) - 1}$$

$$\leq \frac{\sqrt{m}}{m-1} \sum |\lambda| \mathrm{pr}_x(\lambda) + \sum_{|\lambda| > \sqrt{m}} 2c|\lambda|$$

provided that $\sqrt{m}/c \geq 2$. In case (i), this is infinitesimal for $m \simeq \infty$ (so that the series converges), and in case (ii) it is limited. $\square$

Let $x_1, \ldots, x_\nu$ and x be random variables, not necessarily defined over the same finite probability space. We say that the $x_1, \ldots, x_\nu$ are *dominated in distribution by* x in case there is an $a \ll \infty$ such that for all $n = 1, \ldots, \nu$ and all natural numbers k we have

$$\Pr\{k < |x_n| \leq k+1\} \leq a\Pr\{k < |x| \leq k+1\}.$$

This is certainly the case if the $x_1, \ldots, x_\nu$ all have the same probability distribution as x. In the counterexample given earlier in this chapter, $x_3, x_4, \ldots, x_\nu$ are not dominated in distribution by any random variable.

Let $x_1, \ldots, x_\nu$ be independent random variables of mean 0 that are dominated in distribution by the random variable x. If one rereads the proof of Theorem 16.3, one finds that everything extends to this more general case except for the estimate of $\left|\mathbf{E}x_n^{(cn)} - \mathbf{E}x_m^{(cm)}\right|$, which was used in case (ii). Thus we have the following result.

Theorem 16.4 *Let x be a random variable and let $x_1, \ldots, x_\nu$ be independent random variables of mean 0 that are dominated in distribution by x. If x is L^1, then $(x_1 + \cdots + x_n)/n$ converges to 0 a.s.*

Here is a counterexample in case (ii). Let $\nu \simeq \infty$, let μ be such that $\mu/\nu \simeq \infty$, and let x be 1 with probability $1 - 1/\mu$ and $-\mu + 1$ with probability $1/\mu$. Then x is not L^1, but $||x||_1 = 2 - 2/\mu \ll \infty$, and $\mathbf{E}x = 0$. Let $x_1', \ldots, x_\nu'$ be independent observations of x. Then a.s. $x_n' = 1$ for all $n \leq \nu$. Let $x_1, \ldots, x_\nu$ be the same as $x_1', \ldots, x_\nu'$ except that x_1' is replaced by 0, the next 2! random variables x_n' are unchanged, the next 3! random variables are replaced by 0, the next 4! are unchanged, and so forth up to ν. Then the $x_1, \ldots, x_\nu$ are independent random variables of mean 0 that are dominated in distribution by x with $||x||_1 \ll \infty$, but a.s. the $(x_1 + \cdots + x_n)/n$ are of unlimited fluctuation.

The point of the strong law of large numbers is that eventually the averages $(x_1 + \cdots + x_n)/n$ of repeated independent observations of x settle down near $\mathbf{E}x$. This is true if x is L^1 by Theorem 16.3 (i). If we assume only that $||x||_1 \ll \infty$, then the conclusion of Theorem 16.3 (ii) is rather uninformative: it tells us merely that the averages can fluctuate substantially only a limited number of times. Nevertheless, if $||x||_1 \ll \infty$, then by Theorem 6.1 (applied to the sequence $\mathbf{E}|x^{(a)}|$) there is an unlimited a such that $x^{(a)}$ is L^1. If we choose $\mu \leq \nu$ unlimited but sufficiently small, then we can ensure that a.s. $x_n = x_n^{(a)}$ for all $n \simeq \mu$, so that by Theorem 16.3 (i) the sequence $(x_1 + \cdots + x_n)/n$ for $n \leq \mu$ converges a.s. to $\mathbf{E}x^{(a)}$.

We say that a random variable x is *near* L^1 in case there is an L^1 random variable y such that $x \simeq y$ a.e., in which case we call $\mathbf{E}y$ a *reduced expectation* of x. By the Lebesgue theorem, if z is also an L^1 random variable with $z \simeq x$ a.e., then $\mathbf{E}z \simeq \mathbf{E}y$. Thus the reduced expectation of a near L^1 random variable is limited and is uniquely defined up to an infinitesimal.

*Theorem 16.5 *Let x be a random variable. Then the following are equivalent:*

(i) *x is near L^1,*

(ii) *for some $a \simeq \infty$ we have that $\Pr\{|x| > a\} \simeq 0$ and $\mathbf{E}|x^{(a)}| \ll \infty$,*

(iii) *for some $a \simeq \infty$ we have that $\Pr\{|x| > a\} \simeq 0$ and $x^{(a)}$ is L^1.*

Proof. As already remarked, it follows from Theorem 6.1 that (ii) $\Rightarrow$ (iii). The implication (iii) $\Rightarrow$ (i) is obvious. Suppose that (i) holds. Then clearly $\Pr\{|x| > a\} \simeq 0$ for all $a \simeq \infty$. Let y be L^1 and such that $x \simeq y$ a.e., and let K be such that $\mathbf{E}|y| \ll K \ll \infty$. Suppose that $\mathbf{E}|x^{(a)}| \simeq \infty$ for all $a \simeq \infty$. Then the set of all a such that $\mathbf{E}|x^{(a)}| \geq K$ contains all $a \simeq \infty$, and so contains some $a \ll \infty$. But $|x^{(a)}| \leq |x| \simeq |y|$ a.e., and since $x^{(a)}$ is clearly L^1, the Lebesgue theorem implies that $\mathbf{E}|x^{(a)}| \lesssim \mathbf{E}|y| \ll K$, which is a contradiction. $\square$

We have the implications

$$x \text{ is } L^1 \quad \Rightarrow \quad ||x||_1 \ll \infty \quad \Rightarrow \quad x \text{ is near } L^1,$$

with neither of the of the reverse implications holding in general.

The following theorem is a corollary of Theorems 16.3 (i) and 16.5.

*Theorem 16.6 *Let x be near L^1 and let $x_1, \ldots, x_\nu$, with $\nu \simeq \infty$, be independent random variables with the same probability distribution as x. Then there is an unlimited $\mu \leq \nu$ such that the $(x_1 + \cdots + x_n)/n$ for $n = 1, \ldots, \mu$ converge to the reduced expectation of x.*

Chapter 17

Nearly equivalent processes

In Chapter 3 we defined the internal notion of equivalence of two stochastic processes. Now we shall define an external notion: near equivalence. The intuitive content is that two stochastic processes are nearly equivalent in case they cannot be told apart by observations that are incapable of resolving infinitesimals.

Recall that the trajectories of a stochastic process ξ indexed by a finite set T and defined over a finite probability space $\langle \Omega, \mathrm{pr} \rangle$ are elements of the finite subset Λ_ξ of $\mathbf{R}^T$. If Λ is a finite subset of $\mathbf{R}^T$ and $F : \Lambda \to \mathbf{R}$, then we call F a *functional* defined over Λ. If F is a functional defined over Λ and $\Lambda_\xi \subseteq \Lambda$, then we write $F(\xi)$ for the random variable whose value at each ω in Ω is $F(\xi(\omega))$. The space Λ is a metric space with respect to the metric

$$\rho(\lambda, \mu) = \max_{t \in T} |\lambda(t) - \mu(t)|,$$

and a functional is (nearly) *continuous* in case $\rho(\lambda, \mu) \simeq 0$ (which is equivalent to $\lambda(t) \simeq \mu(t)$ for all t) implies $F(\lambda) \simeq F(\mu)$. We say that F is *limited* in case $|F(\lambda)| \ll \infty$ for each λ in Λ. Since $|F|$ attains its maximum somewhere, this is the same as saying that $\max |F| \ll \infty$.

Let ξ and η be two stochastic processes indexed by the same finite set T but defined over possibly different finite probability spaces. Notice that ξ and η are equivalent if and only if $\mathbf{E} F(\xi) = \mathbf{E} F(\eta)$ for all functionals F (defined on $\Lambda_\xi \cup \Lambda_\eta$; but then we must have $\Lambda_\xi = \Lambda_\eta$). We say that ξ and η are *nearly equivalent* in case $\mathbf{E} F(\xi) \simeq \mathbf{E} F(\eta)$ for all limited continuous functionals F on $\Lambda_\xi \cup \Lambda_\eta$.

Theorem 17.1 *Let ξ be a stochastic process defined over $\langle \Omega, \mathrm{pr} \rangle$. Suppose that Ω is also a finite probability space with respect to pr', and that $\sum |\mathrm{pr}(\omega) - \mathrm{pr}'(\omega)| \simeq 0$. Let ξ' be the same function as ξ, but regarded as a*

72

stochastic process on $\langle \Omega, \mathrm{pr}' \rangle$. Then ξ and ξ' are nearly equivalent.

Proof. Let F be a limited functional. Then

$$|\mathbf{E}F(\xi) - \mathbf{E}'F(\xi')| \leq \sum |F(\xi(\omega))| \, |\mathrm{pr}(\omega) - \mathrm{pr}'(\omega)|$$

$$\leq \max |F| \sum |\mathrm{pr}(\omega) - \mathrm{pr}'(\omega)| \simeq 0. \ \square$$

Theorem 17.2 *Let ξ and η be stochastic processes indexed by T and defined over the same finite probability space. If a.s. $\xi(t) \simeq \eta(t)$ for all t, then ξ and η are nearly equivalent.*

Proof. If F is a limited functional, then $F(\xi)$ and $F(\eta)$ are L^1, and if F is continuous, then $F(\xi) \simeq F(\eta)$ a.s., so that by the Lebesgue theorem $\mathbf{E}F(\xi) \simeq \mathbf{E}F(\eta)$. $\square$

In the next theorem, A is any formula, internal or external. This theorem shows that our definition of near equivalence has the desired intuitive content.

Theorem 17.3 *Suppose that for all λ and μ in Λ we have that $\rho(\lambda, \mu) \simeq 0$ implies that $\mathrm{A}(\lambda)$ if and only if $\mathrm{A}(\mu)$. Let ξ and η be nearly equivalent stochastic processes with $\Lambda_\xi \cup \Lambda_\eta \subseteq \Lambda$. Then $\mathrm{A}(\xi)$ a.s. if and only if $\mathrm{A}(\eta)$ a.s.*

Proof. Without loss of generality, we assume that ξ and η are defined over Λ_ξ and Λ_η. Suppose that $\mathrm{A}(\xi)$ a.s. We need to prove that $\mathrm{A}(\eta)$ a.s.

Let $\varepsilon \gg 0$. Then there is a set $\Phi \subseteq \Lambda_\xi$ such that $\mathrm{Pr}_\xi \Phi \geq 1 - \varepsilon$ and $\mathrm{A}(\lambda)$ holds for all λ in Φ. For $\beta > 0$ let

$$F_\beta(\lambda) = \left(1 - \frac{\rho(\lambda, \Phi)}{\beta} \right)^+.$$

Then $0 \leq F_\beta \leq 1$, so F_β is limited. If $\beta \gg 0$, then F_β is continuous. Let $\Phi_\beta = \{\lambda : \rho(\lambda, \Phi) \leq \beta\}$, and note that

$$\chi_{\Phi_\beta} \geq F_\beta \geq \chi_\Phi.$$

Therefore, if $\beta \gg 0$ we have

$$\mathrm{Pr}_\eta \Phi_\beta \geq \mathbf{E}F_\beta(\eta) \simeq \mathbf{E}F_\beta(\xi) \geq \mathrm{Pr}_\xi \Phi \geq 1 - \varepsilon,$$

so that $\mathrm{Pr}_\eta \Phi_\beta \geq 1 - 2\varepsilon$. Since this holds for all $\beta \gg 0$, it holds for some $\beta \simeq 0$ by overspill. But for $\beta \simeq 0$ we have $\mathrm{A}(\lambda)$ for all λ in Φ_β. Since $\varepsilon \gg 0$ is arbitrary, we have $\mathrm{A}(\eta)$ a.s. $\square$

Corollary 1. *Let ξ and η be nearly equivalent stochastic processes and let F be a functional. If F is continuous at a.e. trajectory of ξ, then F is continuous at a.e. trajectory of η.*

Corollary 2. *Let ξ and η be nearly equivalent stochastic processes indexed by a finite subset T of $\mathbf{R}$. If a.e. trajectory of ξ is continuous, then a.e. trajectory of η is continuous.*

Theorem 17.4 *Let ξ and η be nearly equivalent stochastic processes and let F be a limited functional that is continuous at a.e. trajectory of ξ. Then $\mathbf{E}F(\xi) \simeq \mathbf{E}F(\eta)$.*

Proof. Without loss of generality, we assume that ξ and η are defined over Λ_ξ and Λ_η, and that $1 \leq F \leq c \ll \infty$. Let $\varepsilon \gg 0$. By Corollary 1 there is a subset Φ of $\Lambda_\xi \cup \Lambda_\eta$ such that $\mathrm{Pr}_\xi(\Phi \cap \Lambda_\xi) \geq 1-\varepsilon$ and $\mathrm{Pr}_\eta(\Phi \cap \Lambda_\eta) \geq 1-\varepsilon$, and F is continuous at each element of Φ. Now define the functional G as follows: if $\lambda \in \Phi$ let $G(\lambda) = F(\lambda)$, and if $\lambda \notin \Phi$ let

$$G(\lambda) = \frac{\min_{\mu \in \Phi} \rho(\lambda, \mu) F(\mu)}{\rho(\lambda, \Phi)}.$$

For all λ in $\Lambda = \Lambda_\xi \cup \Lambda_\eta$, let λ^* be an element of Φ such that $\rho(\lambda, \Phi) = \rho(\lambda, \lambda^*)$, and let λ' be an element μ of Φ at which $\rho(\lambda, \mu) F(\mu)$ achieves its minimum on Φ. Then $1 \leq G \leq c$. This is obvious for $\lambda \in \Phi$, and for $\lambda \notin \Phi$ we have

$$1 \leq \frac{\rho(\lambda, \lambda') F(\lambda')}{\rho(\lambda, \Phi)} = G(\lambda) \leq \frac{\rho(\lambda, \lambda^*) F(\lambda^*)}{\rho(\lambda, \Phi)} = F(\lambda^*) \leq c. \qquad (17.1)$$

I claim that G is everywhere continuous. To see this, let $\rho(\lambda_1, \lambda_2) \simeq 0$. If $\lambda_1, \lambda_2 \in \Phi$, then $G(\lambda_1) - G(\lambda_2) = F(\lambda_1) - F(\lambda_2) \simeq 0$. If $\lambda_1 \notin \Phi$ but $\rho(\lambda_1, \Phi) \simeq 0$, then $\rho(\lambda_1, \lambda_1') \simeq 0$ by the definition of λ_1', since $1 \leq F$. Therefore $\rho(\lambda_1', \lambda_1^*) \simeq 0$ and so $F(\lambda_1') \simeq F(\lambda_1^*)$. By (17.1),

$$1 \leq \frac{\rho(\lambda_1, \lambda_1')}{\rho(\lambda, \Phi)} \simeq 1$$

and so $G(\lambda_1) \simeq F(\lambda_1^*)$. Since also $G(\lambda_2) \simeq F(\lambda_2^*)$, we have $G(\lambda_1) \simeq G(\lambda_2)$. Finally, if $\rho(\lambda_1, \Phi) \gg 0$, then it is clear that $G(\lambda_1) \simeq G(\lambda_2)$, which establishes the claim.

Therefore $\mathbf{E}G(\xi) \simeq \mathbf{E}G(\eta)$, by the definition of near equivalence. But $\mathbf{E}|F(\xi) - G(\xi)| \leq c\varepsilon$ and $\mathbf{E}|F(\eta) - G(\eta)| \leq c\varepsilon$, and since $\varepsilon \gg 0$ is arbitrary we have $\mathbf{E}F(\xi) \simeq \mathbf{E}F(\eta)$. $\square$

Chapter 18

The de Moivre-Laplace-Lindeberg-Feller-Wiener-Lévy-Doob-Erdös-Kac-Donsker-Prokhorov theorem

Let ξ be a stochastic process indexed by a near interval T (for example, the normalized martingale associated to a series of random variables in Chapter 14). We say that ξ is a (near) *Wiener process* in case ξ is nearly equivalent to the Wiener walk on T. The following is a version of the de Moivre-Laplace central limit theorem that contains Lindeberg's theorem on the sufficiency of his condition, Feller's theorem on its necessity, Wiener's theorem on the continuity of trajectories for his process, the Lévy-Doob characterization of it as the only normalized martingale with continuous trajectories, and the invariance principle of Erdös and Kac as extended by Donsker and Prokhorov.

Theorem 18.1 *Let ξ be a normalized martingale, indexed by a near interval, with $\xi(a) = 0$. Then the following are equivalent:*

(i) *ξ is a Wiener process,*

(ii) *a.s. ξ is continuous for all t, and $\xi(b)$ is L^2,*

(iii) *ξ satisfies the Lindeberg condition.*

    Proof. We will show that (i) $\Rightarrow$ (ii) $\Rightarrow$ (iii) $\Rightarrow$ (i).

Suppose that (i) holds. By the corollary to Theorem 13.1, a.e. trajectory of the Wiener walk is continuous for all t, so by Corollary 2 to Theorem 17.3, a.e. trajectory of ξ is continuous for all t. We need to show that $\xi(b)$ is L^2. Let

$$f_{(c)}(\lambda) = \begin{cases} \lambda^2, & |\lambda| \leq c, \\ c^2, & |\lambda| > c, \end{cases}$$

and observe that a random variable y is L^2 if and only if for all $\varepsilon \gg 0$ there is a $c \ll \infty$ such that $\mathbf{E}f_{(c)}(y) \geq \mathbf{E}y^2 - \varepsilon$. If $c \ll \infty$, then $f_{(c)}$ is limited and continuous, so if we let $F_{(c)}(\xi) = f_{(c)}(\xi(b))$, then $F_{(c)}$ is a limited continuous functional. Let w be the Wiener walk. It is easily seen that $w(b)$ is L^2; for example, because

$$\mathbf{E}w(b)^4 = \mathbf{E}\left(\sum dw(t)\right)^4 = \mathbf{E}\sum dw(t_1)dw(t_2)dw(t_3)dw(t_4)$$

$$= 3\sum_{t \neq s}\mathbf{E}dw(t)^2 dw(s)^2 + \sum dw(t)^4$$

$$= 3\sum_{t \neq s} dt\, ds + \sum dt^2 \simeq 3(b-a)^2 \ll \infty.$$

Therefore, for $\varepsilon \gg 0$ there is a $c \ll \infty$ such that $\mathbf{E}f_{(c)}(w(b)) \geq b - a - \varepsilon$, and so

$$\mathbf{E}f_{(c)}(\xi(b)) \gtrsim b - a - \varepsilon = \mathbf{E}\xi(b)^2 - \varepsilon.$$

Hence $\xi(b)$ is L^2, and thus (i) $\Rightarrow$ (ii).

Suppose that (ii) holds. In the identity

$$\xi(b)^2 = 2\sum \xi(t)d\xi(t) + \sum d\xi(t)^2,$$

the left hand side is L^1 by hypothesis. The next term is L^1 because the expectation of its square is

$$4\mathbf{E}\sum \xi(t)^2 dt = 4\sum(t-a)dt \simeq 2(b-a)^2 \ll \infty.$$

Therefore the last term is L^1, and consequently $\sum d\xi(t)^{(\varepsilon)2}$ is L^1. But if $\varepsilon \gg 0$, then $\sum d\xi(t)^2 = \sum d\xi(t)^{(\varepsilon)2}$ a.s. by the continuity of trajectories. By the Lebesgue theorem, their expectations are infinitely close, and this is the Lindeberg condition. Thus (ii) $\Rightarrow$ (iii).

Suppose that (iii) holds. The idea of the proof that (iii) $\Rightarrow$ (i) is simple. Observe the process ξ at the times t_n as in Fig. 18.1, where ε is a huge infinitesimal. Then it almost surely goes up or down by nearly $\sqrt{\varepsilon}$, and by the martingale property these occur with nearly equal probability. The times t_n are random variables, but since the quadratic variation of the process (which is nearly $n\varepsilon$) is nearly equal to the elapsed time $t - a$, they

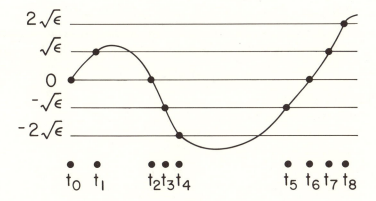

Figure 18.1: A Wiener process

behave as if they were spaced ε apart on the average (this is not indicated in Fig. 18.1), and the process looks like a Wiener process.

We shall prove that (iii) $\Rightarrow$ (i) under the weaker assumption that ξ is *nearly normalized*, by which we mean that a.s. $\tau_\xi(t) \simeq t - a$ for all t. By Theorems 14.3 and 17.2, there is no loss of generality in assuming that for some $\alpha \simeq 0$ we have $|d\xi(t)| \leq \alpha$ for all t and all ω.

Let $\overline{b} - b \simeq \infty$, and let $\overline{T}$ be the union of T and the set of all numbers of the form

$$b + k\frac{\overline{b} - b}{m} \text{ for } k = 1, \ldots, m, \text{ where } \frac{\overline{b} - b}{m} \simeq 0.$$

We extend ξ to $\overline{T}$ by setting $\xi(t) = \xi(b) + \overline{w}(t)$, where $\overline{w}$ is the Wiener walk on $\overline{T} \setminus T$. This extension preserves the properties of ξ assumed above. The point of this extension is to avoid having to worry about when the times t_n become undefined for the original process.

We denote the predecessor of t by $t - d_* t$. We say that ξ *crosses* λ at t in case either $\xi(t - d_* t) < \lambda$ and $\xi(t) \geq \lambda$, or else $\xi(t - d_* t) > \lambda$ and $\xi(t) \leq \lambda$. Let $\varepsilon > 0$ with $\sqrt{\varepsilon} > 2\alpha$, and let $\nu = [(b - a)/\varepsilon]$. Let $t_0 = a$ and define t_n inductively, for $n = 1, \ldots, \nu$, to be the first time subsequent to t_{n-1} at which ξ crosses some $k_n\sqrt{\varepsilon}$, where k_n is an integer with the same parity as n (and if there is no such time, then we define t_n to be $\overline{b}$). See Fig. 18.1.

Let $\widetilde{P}_n$ be the algebra generated by $\xi(t_1), \ldots, \xi(t_n)$. I claim that the $\xi(t_n)$ are a martingale with respect to this filtration. To see this, let A be an event with χ_A in $\widetilde{P}_n$. Then

$$\chi_A (\xi(t_{n+1}) - \xi(t_n)) = \sum \eta(s)d\xi(s)$$

with η a $\tilde{P}$-process: $\eta(s) = 1$ if the event A has occurred and $t_n \leq s < t_{n+1}$, and otherwise $\eta(s) = 0$. Therefore $\mathbf{E}\chi_A(\xi(t_{n+1}) - \xi(t_n)) = 0$, and since this is true for all A with χ_A in $\tilde{P}_n$, it follows that

$$\mathbf{E}_{\tilde{P}_n}(\xi(t_{n+1}) - \xi(t_n)) = 0,$$

which establishes the claim.

Suppose that $0 \ll \varepsilon \ll \infty$. Then a.s.

$$\left| \xi(t_{n+1}) - \xi(t_n) \mp \sqrt{\varepsilon} \right| \leq 2\alpha$$

for all $n < \nu$, since the only way this could fail is for t_{n+1} to be $\overline{b}$—but since $\overline{b} - b \simeq \infty$, and $\nu \ll \infty$ for $\varepsilon \gg 0$, it is easy to see that a.s. this does not occur. Notice also that $|\xi(t_{n+1}) - \xi(t_n)| \leq \sqrt{\varepsilon} + 2\alpha \ll \infty$ everywhere, for all $n < \nu$. Since $\xi(t_{n+1}) - \xi(t_n)$ has mean 0, we have

$$\Pr\left\{ \left| \xi(t_{n+1}) - \xi(t_n) - \sqrt{\varepsilon} \right| \leq 2\alpha \right\} \simeq \frac{1}{2},$$

$$\Pr\left\{ \left| \xi(t_{n+1}) - \xi(t_n) + \sqrt{\varepsilon} \right| \leq 2\alpha \right\} \simeq \frac{1}{2}.$$

The same holds for the conditional probability with respect to $\tilde{P}_n$. Since $\nu \ll \infty$ for $\varepsilon \gg 0$, this means that

$$\sum_\pi \left| \Pr\left\{ \left| \xi(t_{n+1}) - \xi(t_n) - (-1)^{\pi(n)}\sqrt{\varepsilon} \right| \leq 2\alpha \text{ for all } n \right\} - \frac{1}{2^\nu} \right| \simeq 0, \quad (18.1)$$

where the sum is over all 2^ν mappings π of $\{0, \dots, \nu - 1\}$ into $\{0, 1\}$.

Now use overspill. Since the set of all ε for which the left hand side of (18.1) (i.e., everything but the $\simeq 0$) is $\leq \varepsilon$ contains all $\varepsilon \gg 0$, it contains all sufficiently large $\varepsilon \simeq 0$. Hence (18.1) continues to hold for all sufficiently large $\varepsilon \simeq 0$. Now fix ε (the "huge infinitesimal") so that $\varepsilon \simeq 0$, $\alpha/\varepsilon \simeq 0$, and (18.1) holds.

Let the random variable $\nu(t)$ be the largest n with $n \leq \nu$ such that $t_n \leq t$. For t in T, let

$$\varsigma(t) = \sum_{n=1}^{\nu(t)} \left(\xi(t_n) - \xi(t_{n-1}) \right)^2 + \left(\xi(t) - \xi(t_{\nu(t)}) \right)^2. \quad (18.2)$$

Then

$$d\varsigma(t) = d\xi(t)^2 + 2(\xi(t) - \xi(t_{\nu(t)}))d\xi(t),$$

so that $|d\varsigma(t)| \leq \gamma|d\xi(t)|$ for all t in T' and all ω, where $\gamma = \alpha + 2(\sqrt{\varepsilon} + \alpha) \simeq 0$. Consequently,

$$\left\| \sum d\varsigma(t) - \sum \mathbf{E}_t d\varsigma(t) \right\|_2^2 \leq \sum \|d\varsigma(t)\|_2^2 \leq \gamma^2 \sum \|d\xi(t)\|_2^2 \simeq 0.$$

By Theorems 11.1 and 7.1, a.s.

$$\varsigma(t) \simeq \sum_{s<t} \mathbf{E}_s d\varsigma(s) = \sum_{s<t} \mathbf{E}_s d\xi(s)^2 \simeq \tau_\xi(t) \simeq t - a$$

for all t in T. Since $\alpha/\varepsilon \simeq 0$, a.s. each term in the sum in (18.2) is $\sim \varepsilon$. The last term in (18.2) is $\simeq 0$, so that a.s. $\varsigma(t) \simeq \nu(t)\varepsilon$. Thus a.s.

$$\nu(t)\varepsilon \simeq t - a \qquad (18.3)$$

for all t in T.

Let Π be the finite probability space of all mappings π of $\{1, \ldots, \nu\}$ into $\{0, 1\}$, with $\mathrm{pr}(\pi) = 1/2^\nu$ for all π in Π. Let w_ε be the stochastic process indexed by T and defined over $\langle \Pi, \mathrm{pr} \rangle$ by

$$w_\varepsilon(t) = \sum_{n \le (t-a)/\varepsilon} (-1)^{\pi(n)} \sqrt{\varepsilon}.$$

By (18.1) and (18.3), and Theorems 17.1 and 17.2, ξ is nearly equivalent to w_ε. If ξ' is any other process satisfying our hypotheses—in particular, if ξ' is the Wiener walk on T—then for sufficiently large $\varepsilon \simeq 0$, both ξ and ξ' are nearly equivalent to w_ε, and so are nearly equivalent to each other. Therefore ξ is a Wiener process, and thus (iii) $\Rightarrow$ (i). $\square$

This is an arbitrary stopping point. More can be done. I hope that someone will write a truly elementary book on stochastic processes along these lines, complete with exercises and applications.

Appendix

Introduction

The purpose of this appendix is to demonstrate that theorems of the conventional theory of stochastic processes can be derived from their elementary analogues by arguments of the type usually described as generalized nonsense; there is no probabilistic reasoning in this appendix. This shows that the elementary nonstandard theory of stochastic processes can be used to derive conventional results; on the other hand, it shows that neither the elaborate machinery of the conventional theory nor the devices from the full theory of nonstandard analysis, needed to prove the equivalence of the elementary results with their conventional forms, add anything of significance: the elementary theory has the same scientific content as the conventional theory. This is intended as a self-destructing appendix.

We assume a knowledge of conventional measure theory and of nonstandard analysis in the form of Internal Set Theory (IST), for which see [1].[1]

The elementary axioms of Chapter 4 are theorems of IST. Axioms (1) and (2) follow from the transfer principle, axiom (3) follows from the idealization principle, and external induction (4) follows from the standardization principle. The sequence principle (*5) is a particular case of the saturation principle; see [2]. Consequently, all of the theorems of the text are theorems of IST.

Nearby elementary processes

Theorems of conventional mathematics are internal, so to prove such a theorem about a stochastic process we can assume that the process is standard, by transfer. This implies that its indexing set and the probability

---

[1]The references are at the end of this appendix.

space over which it is defined are standard.

Let ξ_0 be a standard stochastic process indexed by T_0 and defined over $\langle \Omega_0, S_0, \mathrm{Pr}_0 \rangle$. By a *nearby elementary process* we mean a stochastic process ξ indexed by a finite subset T of T_0 containing all of the standard elements of T_0, defined over $\langle \Omega_0, S_0, \mathrm{Pr}_0 \rangle$ but taking only finitely many values, such that

$$\sum_{t \in T} |\xi(t) - \xi_0(t)| \simeq 0 \tag{A.1}$$

except on a set of infinitesimal measure, and such that if ξ_0 takes values in L^p, where $1 \leq p \leq \infty$ is standard, then

$$\sum_{t \in T} \|\xi(t) - \xi_0(t)\|_p \simeq 0. \tag{A.2}$$

Observe that if ξ_0 takes values in L^p for some standard p, then (A.1) follows from (A.2), by the Chebyshev inequality. Let S be the σ-algebra generated by the $\xi(t)$ for t in T; then S is a finite Boolean subalgebra of S_0. Let Ω be the set of all atoms of S of strictly positive measure, and define pr on Ω by $\mathrm{pr}(\omega) = \mathrm{Pr}_0(\omega)$. Then $\langle \Omega, \mathrm{pr} \rangle$ is a finite probability space. We can consider ξ either as a stochastic process defined over $\langle \Omega, \mathrm{pr} \rangle$, as in the text, or, for purposes of comparison with ξ_0, as defined over $\langle \Omega_0, S_0, \mathrm{Pr}_0 \rangle$; we will not distinguish notationally between the two notions.

To obtain a nearby elementary process, it suffices to take a sufficiently large unlimited number of terms of the decimal expansion of $\xi_0(t)$ for t in T. For any positive real number x and natural number n, we let $x_{[n]}$ be the largest number $\leq x$ of the form

$$\sum_{k=-n}^{n} a_k 2^{-k},$$

where each a_k is 0 or 1, and for $x < 0$ we let $x_{[n]} = -(-x)_{[n]}$.

Theorem A.1 *Let ξ_0 be a standard stochastic process. Then there exists a nearby elementary process.*

Proof. The existence of T follows from the idealization principle. Let $\varepsilon > 0$ be infinitesimal. Let P be the set of all p in $[1, \infty]$ such that ξ_0 takes values in L^p. If P is non-empty, then it contains an element p_0 such that $p_0 \geq p$ for all standard p in P, by idealization. Then by the Lebesgue dominated convergence theorem, if n is sufficiently large we have

$$\sum_{t \in T} \|\xi_0(t)_{[n]} - \xi_0(t)\|_{p_0} \leq \varepsilon.$$

Let $\xi(t) = \xi_0(t)_{[n]}$ for t in T; then (A.2) holds for all standard p in P. As already remarked, this implies (A.1). If P is empty, let

$$N_n = \left\{ \max_{t \in T} |\xi_0(t)| \geq 2^n \right\}.$$

Choose n so large that $\mathrm{Pr}_0\, N_n \leq \varepsilon$ (this is possible since the N_n decrease to the empty set), n is greater than the cardinality of T, and $n2^{-n} \leq \varepsilon$. Again let $\xi(t) = \xi_0(t)_{[n]}$ for t in T. Then (A.1) holds because $|\xi(t) - \xi_0(t)| \leq 2^{-n}$ except on N_n. $\square$

It is clear that the $\xi(t)$ given by this construction are independent if the $\xi_0(t)$ are. If ξ_0 is a martingale, we must modify the construction a bit to obtain an elementary martingale.

Theorem A.2 *Let ξ_0 be a standard stochastic process, indexed by a subset T_0 of $\mathbf{R}$ and taking values in L^1, that is a supermartingale, submartingale, or martingale. Then there exists a nearby elementary process with the same property.*

Proof. Choose a finite subset T of T_0 containing all of its standard elements, and use the notation $(T', dt, a, \text{etc.})$ of the text. We suppose that ξ_0 is adapted to a certain filtration P_0. We will choose an n and let P_t be the finite Boolean algebra generated by the $\xi_0(s)_{[n]}$ for $s \leq t$, $s \in T$, so that $P_t \subseteq P_{0t}$. Then we define ξ by

$$d\xi(t) = d\left(\xi_0(t)_{[n]}\right) + \mathbf{E}\left\{d\xi_0(t) - d\left(\xi_0(t)_{[n]}\right) \mid P_t\right\}, \quad t \in T',$$

$$\xi(t) = \xi_0(a)_{[n]} + \sum_{s < t} d\xi(s), \quad t \in T.$$

Then

$$\mathbf{E}\left\{d\xi(t) \mid P_t\right\} = \mathbf{E}\left\{d\xi_0(t) \mid P_t\right\}.$$

Since $P_t \subseteq P_{0t}$, this has the right sign; that is, if ξ_0 is a supermartingale, submartingale, or martingale, then so is ξ. Let $\varepsilon > 0$ be infinitesimal, and let p_0 be as in the previous proof. If n is sufficiently large, then

$$\sum_{t \in T} \|\xi(t) - \xi_0(t)\|_{p_0} \leq \varepsilon$$

by the Lebesgue dominated convergence theorem. $\square$

These proofs show that the infinitesimal implicit in the definition of a nearby elementary process can be chosen to be arbitrarily small.

Equivalence of analytical properties

A typical theorem of the theory of stochastic processes asserts that under certain analytical hypotheses, certain probabilistic conclusions hold almost surely for the sample paths. To derive a conventional theorem from its elementary analogue, we need to show that the internal analytical hypotheses imply the analogous external analytical hypotheses, and that the external probabilistic conclusion implies the corresponding internal probabilistic conclusion. But to satisfy ourselves that the two formulations are saying the same thing in different languages, we should establish the equivalence of the two forms of the hypotheses and of the two forms of the conclusions.

In this section we will verify that certain internal analytical properties of a standard stochastic process are equivalent to corresponding external analytical properties of a nearby elementary process. The following conventions are in force: ξ_0 is a standard stochastic process, indexed by T_0 and defined over $\langle \Omega_0, S_0, \mathrm{Pr}_0 \rangle$; ξ is a nearby elementary process, indexed by T and defined over $\langle \Omega, \mathrm{pr} \rangle$ as above (and is a supermartingale, etc., if ξ_0 is); a and b are the first and last elements of T, $T' = T \setminus \{b\}$, for t in T' its successor is $t + dt$, and for any function f on T we let $df(t) = f(t + dt) - f(t)$ for t in T'; if $T_0 = \mathbf{N}^+ = \mathbf{N} \setminus \{0\}$, then we choose T to be $\{1, \ldots, \nu\}$ where ν is an unlimited natural number. Notice that if T_0 is the closed interval $[a, b]$, then T is a near interval with first element a and last element b.

Theorem A.3 *Let ξ_0 take values in L^1. If $T_0 = \mathbf{N}^+$, then*

(1) $\sum_{n=1}^{\infty} \|\xi_0(n)\|_1$ *converges* $\iff$ $\sum_{n=1}^{\nu} \|\xi(n)\|_1$ *(nearly) converges,*

(2) $\xi_0(n)$ *converges in L^1* $\iff$ $\xi(n)$ *(nearly) converges in $\mathbf{L}^1$.*

If t_0 is a standard point of T_0, and $T_0 \subseteq \mathbf{R}$, then

(3) ξ_0 *is continuous at t_0 in L^1* $\iff$ ξ *is (nearly) continuous at t_0 in $\mathbf{L}^1$.*

If $T_0 \subseteq \mathbf{R}$, then

(4) ξ_0 *is of bounded variation in L^1* $\iff$ $\sum_{t \in T} \|d\xi(t)\|_1 \ll \infty$.

If T_0 is a closed interval $[a, b]$, then

(5) ξ_0 *is in $L^1(T_0 \times \Omega_0)$* $\iff$ ξ *is L^1 on $T' \times \Omega$.*

(6) ξ_0 *is absolutely continuous in L^1* $\iff$ $d\xi/dt$ *is L^1 on $T' \times \Omega$, for some nearby elementary process ξ.*

Proof. The conclusion of (1) is equivalent to

$$\sum_{n=1}^{\nu} ||\xi_0(n)||_1 \text{ (nearly) converges,} \tag{A.3}$$

since ξ is a nearby elementary process. Thus (1) asserts that a standard series converges if and only if its partial sum up to an unlimited ν nearly converges, and this is easy to see. The proofs of (2) and (3) are entirely similar.

Suppose that the hypothesis of (4) holds. By transfer, there is a standard number K such that for all finite subsets of T_0, and for T in particular, we have $\sum_{t \in T'} ||d\xi_0(t)||_1 \le K$. Therefore the conclusion of (4) holds. Conversely, suppose that the conclusion of (4) holds. Since T contains all standard elements of T_0, there is a fixed standard bound on the variation of ξ_0 for any standard finite subset of T_0, and so by transfer the hypothesis of (4) holds.

The hypothesis of (5) is equivalent to the assertion that

$$||\xi_0^{(n)} - \xi_0^{(m)}||_1 \to 0 \text{ as } n, m \to \infty. \tag{A.4}$$

Since ξ_0 is standard, (A.4) is equivalent to $||\xi_0^{(n)} - \xi_0^{(m)}||_1 \simeq 0$ for $n, m \simeq \infty$, and since ξ is a nearby elementary process, this is equivalent to the conclusion of (5).

For any finite collection I of non-overlapping subintervals of $[a, b]$, let $|I|$ be its total length, and let $\text{var}_0(I)$ be $\sum_i ||\xi_0(b_i) - \xi_0(a_i)||_1$, where the $[a_i, b_i]$ are the intervals of I. Then the hypothesis of (6) is

$$\forall \varepsilon \exists \delta \forall I \left(|I| \le \delta \ \Rightarrow \ \text{var}_0(I) \le \varepsilon \right). \tag{A.5}$$

But this is equivalent to

$$\forall I \left(|I| \simeq 0 \ \Rightarrow \ \text{var}_0(I) \simeq 0 \right); \tag{A.6}$$

to see this, apply the reduction algorithm to (A.6), as follows. Write (A.6) as

$$\forall I \left(\forall^{\text{st}} \delta \ |I| \le \delta \ \Rightarrow \ \forall^{\text{st}} \varepsilon \ \text{var}_0(I) \le \varepsilon \right),$$

rewrite this as

$$\forall^{\text{st}} \varepsilon \forall I \exists^{\text{st}} \delta \left(|I| \le \delta \ \Rightarrow \ \text{var}_0(I) \le \varepsilon \right),$$

use idealization to put this in the form

$$\forall^{\text{st}} \varepsilon \exists^{\text{stfin}} \delta' \forall I \exists \delta \in \delta' \left(|I| \le \delta \ \Rightarrow \ \text{var}_0(I) \le \varepsilon \right),$$

then choose the smallest δ in δ' and apply transfer to obtain (A.5).

Let us call I *good* in case the endpoints of the intervals in I are contained in T. I claim that (A.6) is equivalent to

$$\forall^{\text{good}} I \left(|I| \simeq 0 \;\Rightarrow\; \text{var}_0(I) \simeq 0 \right). \tag{A.7}$$

Clearly, (A.6) implies (A.7). If we apply the reduction algorithm to (A.7), we find that it is equivalent to

$$\forall^{\text{st}} \varepsilon \exists^{\text{st}} \delta \forall^{\text{good}} I \left(|I| \leq \delta \;\Rightarrow\; \text{var}_0(I) \leq \varepsilon \right).$$

Since T contains all of the standard points of $[a,b]$, this implies

$$\forall^{\text{st}} \varepsilon \exists^{\text{st}} \delta \forall^{\text{st}} I \left(|I| \leq \delta \;\Rightarrow\; \text{var}_0(I) \leq \varepsilon \right),$$

which is equivalent by transfer to (A.5), and so to (A.6). This proves the claim.

Thus (A.5) and (A.7) are equivalent. For a good I, let var(I) be $\sum_i \|\xi(b_i) - \xi(a_i)\|_1$. Since ξ is a nearby process, (A.7) is equivalent to

$$\forall^{\text{good}} I (|I| \simeq 0 \;\Rightarrow\; \text{var}(I) \simeq 0).$$

But this asserts that for sets of the form $M = I \times \Omega$, if M is of infinitesimal probability in $T' \times \Omega$ then

$$\mathbf{E}\frac{d\xi}{dt}\chi_M \simeq 0,$$

and this holds if $d\xi/dt$ is L^1 on $T' \times \Omega$, by Theorem 8.1.

For the converse direction, remark that the hypothesis of (6) is equivalent to the existence of a standard function η_0 in $L^1 (T_0 \times \Omega_0)$ such that for all t in T_0,

$$\xi_0(t, \omega_0) = \xi_0(a, \omega_0) + \int_a^t \eta_0(s, \omega_0)ds$$

for a.e. ω_0 in Ω_0. Then let η be a nearby elementary process to η_0, so that η is L^1 on $T' \times \Omega$ by (5). Choose a nearby elementary random variable $\xi(a)$ to $\xi_0(a)$ (that is, choose a nearby elementary process to the standard stochastic process whose index set consists of the single point a and whose random variable is $\xi_0(a)$), and define ξ by

$$\xi(t) = \xi(a) + \sum_{a < s \leq t} \eta(s);$$

then it is easy to verify that ξ is a nearby elementary process to ξ_0, and $d\xi/dt = \eta$ is L^1 on $T' \times \Omega$. $\square$

Theorem A.4 *Let ξ_0 take values in L^1, and suppose that $T_0 \subseteq \mathbf{R}$. If ξ_0 is a martingale or a positive submartingale, then $\|\xi(a)\|_1 \ll \infty$, and we* *have* $||\xi(b)||_1 \ll \infty$ *if and only if* $t \mapsto ||\xi_0(t)||_1$ *is bounded. If* ξ_0 *is a positive supermartingale, then* $||\xi(b)||_1 \ll \infty$, *and we have* $||\xi(a)||_1 \ll \infty$ *if and only if* $t \mapsto ||\xi_0(t)||_1$ *is bounded.*

Proof. Under the first hypothesis, $||\xi_0(t)||_1$ and $||\xi(t)||_1$ are increasing in t, so $||\xi(a)||_1 \ll \infty$. If $||\xi(b)||_1 \ll \infty$, there is a standard K such that $||\xi(t)||_1 \leq K$ for all t in T, so that $||\xi_0(t)||_1 \leq K + 1$ for all standard t and hence, by transfer, for all t. Conversely, if $t \mapsto ||\xi_0(t)||_1$ is bounded, then it has a standard bound K, so that $||\xi_0(b)||_1 \leq K$ and $||\xi(b)||_1 \ll \infty$. The proof under the second hypothesis (when the norms are decreasing) is entirely similar. $\square$

Regular probability measures

Let η_0 be a stochastic process indexed by the set T_0. When T_0 is uncountable, there are many measure-theoretic complications in the theory. To avoid most of these complications, we will always consider the *canonical version* ξ_0 of the process; see [3]. This is a process, indexed by the same set T_0, that is equivalent to η_0; that is, it has the same finite joint distributions. It is defined over path space

$$\Omega_0 = \prod_{t \in T_0} \dot{\mathbf{R}},$$

where $\dot{\mathbf{R}}$ is the one-point compactification of $\mathbf{R}$. Then Ω_0 is a compact Hausdorff space in the product topology. We let S_0 be the σ-algebra of all Borel sets in Ω_0 and we let Pr_0 be the unique regular probability measure such that the stochastic process ξ_0 defined over $\langle \Omega_0, S_0, \mathrm{Pr}_0 \rangle$ by the equation $\xi_0(t, \omega_0) = \omega_0(t)$, for t in T_0 and ω_0 in Ω_0, has the same finite joint distributions as η_0. If η_0 is standard, so is ξ_0.

One advantage of the canonical version is that many interesting subsets of path space are Borel sets, and so are automatically measurable.

We will need two internal results about regular probability measures. Recall that a *directed set* is a set D together with a transitive binary relation $\prec$ on D such that every finite subset of D has an upper bound, and that a *net* is a function $F \mapsto \Phi_F$ defined on D. If it is set-valued, then it is called *increasing* in case $F_1 \prec F_2$ implies $\Phi_{F_1} \subseteq \Phi_{F_2}$, and *decreasing* in case $F_1 \prec F_2$ implies $\Phi_{F_1} \supseteq \Phi_{F_2}$. The point of the following theorem is that it applies to uncountable families of sets.

Theorem A.5 *Let* Pr_0 *be a regular probability measure on a compact Hausdorff space* Ω_0. *Let* $F \mapsto \Phi_F$ *be a decreasing net of closed subsets of* Ω_0, *and let $G \mapsto \Gamma_G$ be an increasing net of open subsets of Ω_0. Then*

$$\text{Pr}_0 \bigcap_F \Phi_F = \inf_F \text{Pr}_0 \Phi_F, \qquad (A.8)$$

$$\text{Pr}_0 \bigcup_G \Gamma_G = \sup_G \text{Pr}_0 \Gamma_G. \qquad (A.9)$$

Proof. Let $\varepsilon > 0$. By definition of regularity, there is a compact set Φ contained in $\bigcup_G \Gamma_G$ such that

$$\text{Pr}_0 \left(\bigcup_G \Gamma_G \setminus \Phi \right) \leq \varepsilon.$$

Since Φ is compact, there is a finite set $\{G_1, \dots, G_n\}$ such that

$$\bigcup_{i=1}^{n} \Gamma_{G_i} \supseteq \Phi.$$

Let G_0 be an upper bound for this finite set. Then $\Gamma_{G_0} \supseteq \Phi$, so that

$$\text{Pr}_0 \left(\bigcup_G \Gamma_G \setminus \Gamma_{G_0} \right) \leq \varepsilon.$$

Since ε is arbitrary, this proves (A.9), and (A.8) follows by duality. $\square$

Theorem A.6 *Let Pr_0 be a regular probability measure on a compact Hausdorff space Ω_0. Let the Φ_{jkF} be closed subsets of Ω_0 and let the Γ_{kjG} be open subsets of Ω_0, where j and k range over $\mathbf{N}$ and F and G range over a directed set $\mathcal{D}$, and let them be decreasing in j and F and increasing in k and G. Then*

$$\text{Pr}_0 \bigcap_j \bigcup_k \bigcap_F \Phi_{jkF} = \sup_{\tilde{k}} \inf_{j,F} \text{Pr}_0 \Phi_{j\tilde{k}(j)F}, \qquad (A.10)$$

where $\tilde{k}$ ranges over all functions from $\mathbf{N}$ to $\mathbf{N}$, and

$$\text{Pr}_0 \bigcup_k \bigcap_j \bigcup_G \Gamma_{kjG} = \sup_{k,\tilde{G}} \inf_j \text{Pr}_0 \Gamma_{kj\tilde{G}(j)}, \qquad (A.11)$$

where $\tilde{G}$ ranges over all functions from $\mathbf{N}$ to $\mathcal{D}$.

Proof. Let

$$p = \text{Pr}_0 \bigcap_j \bigcup_k \bigcap_F \Phi_{jkF}$$

and let $\varepsilon > 0$. Then

$$\forall j \exists k \, \text{Pr}_0 \bigcap_F \Phi_{jkF} \geq p - \varepsilon,$$

so there exists a function $\tilde{k}$ such that

$$\forall j \, \mathrm{Pr}_0 \bigcap_F \Phi_{j\tilde{k}(j)F} \geq p - \varepsilon.$$

By the previous theorem,

$$\inf_{j,F} \mathrm{Pr}_0 \Phi_{j\tilde{k}(j)F} \geq p - \varepsilon.$$

Since ε is arbitrary, this proves the inequality $\leq$ in (A.10). But since

$$\bigcap_j \bigcup_k \bigcap_F \Phi_{jkF} = \bigcup_{\tilde{k}} \bigcap_{j,F} \Phi_{j\tilde{k}(j)F},$$

the reverse inequality is trivial.

Now let

$$p = \mathrm{Pr}_0 \bigcup_k \bigcap_j \bigcup_G \Gamma_{kjG}$$

and let $\varepsilon > 0$. Then

$$\exists k \forall j \exists G \, \mathrm{Pr}_0 \Gamma_{kjG} \geq p - \varepsilon$$

by the previous theorem, so there exist k and $\tilde{G}$ such that for all j,

$$\mathrm{Pr}_0 \Gamma_{kj\tilde{G}(j)} \geq p - \varepsilon.$$

Since ε is arbitrary, this proves the inequality $\leq$ in (A.11), and again the reverse inequality is trivial. $\square$

Equivalence of probabilistic properties

Let $\xi_0 : T_0 \to \mathbf{R}$ be a standard function. The notion of a nearby elementary function ξ is clear: just restrict the definition of a nearby elementary process to the case that the underlying probability space consists of a single point. We frequently have a pair of properties, an internal property A and an external property B, such that A holds for ξ_0 if and only if B holds for ξ. An example is that A is continuity at a standard point t_0 and that B is near continuity at t_0. Notice that if ξ_0 is not standard, the equivalence need not hold. Now let ξ_0 be a standard stochastic process and let ξ be a nearby elementary process, and ask whether A holding a.s. for ξ_0 is equivalent to B holding a.s. for ξ. First we must make sure that the set of all points in the underlying probability space for which A holds is a measurable set;

otherwise the question makes no sense. Even then we cannot argue path-by-path. There is a finite set containing all of the standard paths, and for most interesting stochastic processes any finite set of paths has probability 0 (that is, in general the sample paths of a standard stochastic process are nonstandard!), so that the equivalence of A for a standard function with B for a nearby elementary function does not answer our question.

Theorem A.7 *Let ξ_0 be a standard stochastic process in the canonical version. If $T_0 = \mathbf{N}^+$, then*

(1) $\sum_{n=1}^{\infty} |\xi_0(n)| < \infty$ *a.s.* $\iff$ $\sum_{n=1}^{\nu} |\xi(n)| \ll \infty$ *a.s.,*

(2) $\sum_{n=1}^{\infty} |\xi_0(n)| < \infty$ *a.s.* $\iff$ $\sum_{n=1}^{\nu} |\xi(n)|$ *(nearly) converges a.s.*

If t_0 is a standard point of T_0, and $T_0 \subseteq \mathbf{R}$, then

(3) ξ_0 *is continuous at t_0 a.s.* $\iff$ ξ *is (nearly) continuous at t_0 a.s.*

If T_0 is a compact subset of $\mathbf{R}$, then

(4) ξ_0 *is continuous a.s.* $\iff$ ξ *is (nearly) continuous a.s.*

(5) ξ_0 *has no discontinuities of the second kind a.s.* $\iff$ ξ *is of limited fluctuation a.s.*

If $T_0 \subseteq \mathbf{R}$, then

(6) ξ_0 *is of bounded variation a.s.* $\iff$ ξ *is of limited variation a.s.*

For any T_0,

(7) $\xi_0 = 0$ *a.s.* $\iff$ $\xi \simeq 0$ *a.s.*

Proof. For $i = 1, \ldots, 7$, the equivalence (i) is of the form $A_i \leftrightarrow B_i$, where A_i is an internal formula and B_i is an external formula.

By definition, there is a subset N of path space Ω_0 with $\mathrm{Pr}_0\, N \simeq 0$ such that on N^c we have

$$\sum_{t \in T} |\xi_0(t) - \xi(t)| \simeq 0.$$

Suppose it is false that A_1 a.s., and let

$$\Phi_{jk}^1 = \left\{ \sum_{n=1}^{k} |\xi_0(n)| \geq j \right\},$$

$$\Phi^1 = \bigcap_j \bigcup_k \Phi_{jk}^1.$$

Then $\mathrm{Pr}_0 \, \Phi^1 > 0$, so by transfer there exists a standard $\varepsilon > 0$ such that we have $\mathrm{Pr}_0 \, \Phi^1 > \varepsilon$. By Theorem A.6 and transfer, there is a standard function $\tilde{k}$ such that $\mathrm{Pr}_0 \, \Phi^1_{\frac{1}{\tilde{k}}} > \varepsilon$, where

$$\Phi^1_{\frac{1}{\tilde{k}}} = \bigcap_j \Phi^1_{\frac{1}{j\tilde{k}(j)}}.$$

If j is standard, then $\tilde{k}(j)$ is standard and so $\leq \nu$, and hence

$$\sum_{n=1}^{\nu} |\xi_0(n)| \simeq \infty$$

on $\Phi^1_{\frac{1}{\tilde{k}}}$, since it is bigger than every standard j. Therefore

$$\sum_{n=1}^{\nu} |\xi(n)| \simeq \infty$$

on $N^c \cap \Phi^1_{\frac{1}{\tilde{k}}}$, and $\mathrm{Pr}_0(N^c \cap \Phi^1_{\frac{1}{\tilde{k}}}) > \varepsilon$. Since this is true for some standard $\varepsilon > 0$, it is false that B_1 a.s. Thus B_1 a.s. implies A_1 a.s.

A fortiori, B_2 a.s. implies A_2 (which is identical with A_1) a.s.

Suppose it is false that A_3 a.s., and for G a finite subset of T_0 let

$$\Gamma^3_{kjG} = \bigcup_{\substack{s \in G \\ |s - t_0| \leq 1/j}} \left\{ |\xi_0(s) - \xi_0(t_0)| > \frac{1}{k} \right\},$$

$$\Gamma^3 = \bigcup_k \bigcap_j \bigcup_G \Gamma^3_{kjG}.$$

Then Γ^3 is the set of paths that are discontinuous at t_0. Notice that it is a Borel set, since the uncountable union over G is a union of open sets. Then there is a standard $\varepsilon > 0$ such that $\mathrm{Pr}_0 \, \Gamma^3 > \varepsilon$. By Theorem A.6 and transfer, there exist standard k and $\tilde{G}$ such that $\mathrm{Pr}_0 \, \Gamma^3_{k\tilde{G}} > \varepsilon$, where

$$\Gamma^3_{k\tilde{G}} = \bigcap_j \Gamma^3_{kj\tilde{G}(j)}.$$

If j is standard, then $\tilde{G}(j)$ is a standard finite set, so that each element of it is standard and hence $\tilde{G}(j) \subseteq T$. On $\Gamma^3_{k\tilde{G}}$, therefore, for all standard j there is an s in T with $|s - t_0| \leq 1/j$ and $|\xi_0(s) - \xi_0(t_0)| > 1/k$, so by overspill there is an s in T with $s \simeq t_0$ and $|\xi_0(s) - \xi_0(t_0)| > 1/k$. Consequently, on $N^c \cap \Gamma^3_{k\tilde{G}}$ the process ξ is not nearly continuous at t_0, and so it is false that B_3 a.s. Thus B_3 a.s. implies A_3 a.s.

The proof that B_4 a.s. implies A_4 a.s. is entirely similar: just replace the fixed t_0 by a varying t.

The proof that B_5 a.s. implies A_5 a.s. is similar. For G a finite subset of T_0 consisting of the j points $t_1 < \cdots < t_j$, let

$$\Gamma^5_{kjG} = \bigcap_{i=1}^{j-1} \left\{ |\xi_0(t_i) - \xi_0(t_{i+1})| > \frac{1}{k} \right\},$$

$$\Gamma^5 = \bigcup_k \bigcap_j \bigcup_G \Gamma^5_{kjG},$$

and argue as before. Notice that the Borel set Γ^5 is the set of paths having a discontinuity of the second kind.

The proof that B_6 a.s. implies A_6 a.s. is also similar. For G a finite subset of T_0, let G' be G with its last element deleted, and for t in G', let $t + dt$ be its successor in G. Let

$$\Gamma^6_{jG} = \left\{ \sum_{t \in G'} |\xi_0(t + dt) - \xi_0(t)| > j \right\},$$

$$\Gamma^6 = \bigcap_j \bigcup_G \Gamma^6_{kjG},$$

and argue as before.

Suppose it is false that A_7 a.s. For G a finite subset of T_0, let

$$\Gamma^7_{kG} = \left\{ \max_{t \in G} |\xi_0(t)| > \frac{1}{k} \right\},$$

$$\Gamma^7 = \bigcup_k \bigcup_G \Gamma^7_{kG}.$$

Then there is a standard $\varepsilon > 0$ such that $\mathrm{Pr}_0\, \Gamma^7 > \varepsilon$, and so a standard k and G such that $\mathrm{Pr}_0\, \Gamma^7_{kG} > \varepsilon$. Then $G \subseteq T$, and on Γ^7_{kG} the process ξ is not infinitesimal on T, so it is false that B_7 a.s. Thus B_7 a.s. implies A_7 a.s.

Now let us consider the converse direction.

Suppose that A_2 a.s., and let

$$\Phi^2_{jk} = \left\{ \sum_{n=k}^{\infty} |\xi_0(n)| \leq \frac{1}{j} \right\},$$

$$\Phi^2 = \bigcap_j \bigcup_k \Phi^2_{jk}.$$

Then $\mathrm{Pr}_0 \, \Phi^2 = 1$. Let $\varepsilon > 0$ be standard. By Theorem A.6 and transfer, there is a standard $\tilde{k}$ such that $\mathrm{Pr}_0 \, \Phi_{\tilde{k}}^2 \geq 1 - \varepsilon$, where

$$\Phi_{\tilde{k}}^2 = \bigcap_j \Phi_{j\tilde{k}(j)}^2.$$

On $\Phi_{\tilde{k}}^2$, for all standard j we have

$$\sum_{n=\tilde{k}(j)}^{\infty} |\xi_0(n)| \leq \frac{1}{j},$$

and so on $N^c \cap \Phi_{\tilde{k}}^2$ we have

$$\sum_{n=\tilde{k}(j)}^{\nu} |\xi(n)| \lesssim \frac{1}{j}.$$

Since $\tilde{k}(j)$ is standard for all standard j, for all unlimited $\mu \leq \nu$ we have

$$\sum_{n=\mu}^{\nu} |\xi(n)| \simeq 0,$$

and so $\sum_{n=1}^{\nu} |\xi(n)|$ nearly converges. Hence B_2 a.s., and thus A_2 a.s. implies B_2 a.s.

A fortiori, A_1 a.s. implies B_1 a.s.

Suppose that A_3 a.s., and let

$$\Phi_{jk}^3 = \bigcap_{|s-t_0| \leq 1/k} \left\{ |\xi_0(s) - \xi_0(t_0)| \leq \frac{1}{j} \right\},$$

$$\Phi^3 = \bigcap_j \bigcup_k \Phi_{jk}^3.$$

Then $\mathrm{Pr}_0 \, \Phi^3 = 1$. Let $\varepsilon > 0$ be standard. Then there is a standard $\tilde{k}$ such that $\mathrm{Pr}_0 \, \Phi_{\tilde{k}}^3 \geq 1 - \varepsilon$, where

$$\Phi_{\tilde{k}}^3 = \bigcap_j \Phi_{j\tilde{k}(j)}^3.$$

On $N^c \cap \Phi_{\tilde{k}}^3$ the process ξ is nearly continuous at t_0, and thus A_3 a.s. implies B_3 a.s.

The remaining cases are similar. $\square$

References

[1] Edward Nelson, *Internal set theory: A new approach to nonstandard analysis*, Bull. Amer. Math. Soc. 83 (1977), 1165–1198.

[2] —, *The syntax of nonstandard analysis*, Ann. of Pure and Appl. Logic, to appear.

[3] —, *Regular probability measures on function space*, Ann. of Math. 69 (1959), 630–643.

Index

a, 20
$^{(a)}$, 30
$\mathcal{A}$, 6
absolutely continuous, 24
adapted to, 33
admits k ε-fluctuations, 21
a.e., 25
algebra, 6
almost everywhere, 25
almost surely, 25
a.s., 25
associated martingale, 35
associated normalized martingale, 57
associated predictable process, 34
asymptotic to, 17
at($\mathcal{A}$), 8
atom, 6

b, 20
Borel-Cantelli lemmas, 27

^{c}, 3
canonical version, 86
Cantor's diagonal argument, 46
central limit theorem, 75
Chebyshev inequality, 5
conditional expectation, 7
conditional mean, 7
conditional probability, 8
conjugate exponent, 4
continuous, 23
continuous at t, 23
continuous at t in $\mathbf{L}^p$, 43
continuous functional, 72
convergent, 20, 22

converges in probability, 26
converges to, 20
correlation coefficient, 4
covariance, 4

defined over, 10
de Moivre, 75
disaster, 26
discontinuous at t in $\mathbf{L}^p$, 43
dominated in distribution, 69
Donsker, 75
Doob, 75
dt, 20
$D\xi$, 34
$d\hat{\xi}$, 34
$d\xi(t)$, 20

$\mathbf{E}$, 3
$\mathbf{E}_{\mathcal{A}}$, 7
$\mathbf{E}\{\ |\mathcal{A}\}$, 8
$\mathbf{E}'_{\mathcal{A}}$, 9
equivalent, 10
Erdös, 75
$\mathbf{E}_t$, 33
$\mathbf{E}_T$, 37
event, 3
$\mathbf{E}\{\ |x_1,\ldots,x_n\}$, 8
expectation, 3
external, 13
external induction, 14
external least number principle, 21
ε-continuous, 45
ε-discontinuity, 45
ε-discontinuous, 45
ε-jump, 56

fair game, 36
Feller, 75
filtration, 33
finite probability space, 3
fixed point of discontinuity, 44
Fubini theorem, 32
functional, 72

Gödel, 12

Hölder's inequality, 4

illegal set formation, 13
independent, 11
indexed by, 10
indicator function, 3
indices of k ε-fluctuations, 21
induction theorem, 13
infinitely close, 16
infinitesimal, 16
internal, 13

Jensen's inequality, 5
jump, 56
jump discontinuity, 56

Kac, 75

L^1, 30
L^∞, 31
L^p, 31
$\mathbf{L}^p$, 43
Laplace, 75
Lebesgue theorem, 31
Lévy, 75
limited, 16
limited fluctuation, 21, 22
limited functional, 72
limited variation, 24
Lindeberg, 75
Lindeberg condition, 57

martingale, 33
mean, 3
Minkowski's inequality, 5

N, 13
near, 19
nearby elementary process, 88
near interval, 20
near L^1, 70
nearly, 20
nearly equal, 16
nearly equivalent, 72
nearly normalized, 77
non-degenerate, 36
nonstandard, 12
normalized, 36
normalized martingale, 36

overspill, 18

P, 33
p', 4
Poisson walk, 35
P-process, 33
pr, 3
Pr, 3
pr_A, 7
Pr_A, 8
pr'_A, 8
Pr'_A, 9
predictable process, 34
probability, 3
probability distribution, 10
Prokhorov, 75
proper time, 53
proper time duration, 53
pr_T, 37
P_t, 33

quadratic variation, 58
q_ξ, 58

$\mathbf{R}$, 3
$\overline{\mathbf{R}}$, 16
Radon-Nikodym theorem, 30
random variable, 3
reduced expectation, 70
relativization, 7

Robinson, 15
Robinson's lemma, 19

sample path, 10
sequence principle, 14
standard, 13
standard deviation, 4
stochastic process, 10
strongly greater than, 16
strongly less than, 16
strongly positive, 16
submartingale, 33
supermartingale, 33

T', 20
total variation, 24
trajectory, 10
trend, 34
truncated, 30

unlimited, 16
upcrossings, 50

Var, 4
variance, 4
variation, 24

weakly greater than, 16
weakly less than, 16
Wiener, 75
Wiener process, 75
Wiener walk, 35

χ, 3
$\| \ \|_2$, 3
$\| \ \|_p$, 4
$\| \ \|_\infty$, 4
$\{ \ \}$, 5
$*$, 14
$\simeq$, 16
$\lesssim$, 16
$\gtrsim$, 16
$\ll$, 16
$\gg$, 16
∞, 16
$\simeq \infty$, 16
$\ll \infty$, 16
$-\infty \ll$, 16
$\sim$, 17
σ_ξ^2, 34
ρ, 72

Library of Congress Cataloging-in-Publication Data

Nelson, Edward, 1932-
  Radically elementary probability theory.

  (Annals of mathematics studies ; no. 117)
  Includes index.
  1. Martingales. 2. Stochastic processes.
3. Probabilities. I. Title. II. Series.
QA274.5.N45 1987   519.2   87-3160
ISBN 0-691-08473-4
ISBN 0-691-98474-2 (pbk.)

Edward Nelson is Professor of Mathematics at Princeton University